Standing on the Shoulders of Darwin and Mendel

BOOKS BY DAVID J. GALTON

Molecular genetics of common metabolic disease

Hyperlipidaemia in practice (with W Krone)

The human adipose cell: a model for errors in metabolic regulation

Diabetes mellitus: aetiology and metabolic aspects (co-editor)

Lipids and cardiovascular disease (co-editor)

DNA polymorphisms as disease markers. (co-editor)

In our own image.

Eugenics: the future of human life in the 21st century.

Standing on the Shoulders of Darwin and Mendel

Early Views of Inheritance

by

David J. Galton

CRC Press
Taylor & Francis Group
Boca Raton London New York

CRC Press is an imprint of the
Taylor & Francis Group, an **informa** business

CRC Press
Taylor & Francis Group
6000 Broken Sound Parkway NW, Suite 300
Boca Raton, FL 33487-2742

© 2018 by David J. Galton

No claim to original U.S. Government works

Printed on acid-free paper

International Standard Book Number-13: 978-1-1380-6216-0 (Paperback)
International Standard Book Number-13: 978-1-1380-6217-7 (Hardback)

Library of Congress Cataloging-in-Publication Data

Names: Galton, David J., author.
Title: Standing on the shoulders of Darwin and Mendel : early views of inheritance / David J. Galton.
Description: Boca Raton : Taylor & Francis, 2018. | Includes bibliographical references and index.
Identifiers: LCCN 2017022246 | ISBN 9781138062160 (pbk. : alk. paper) | ISBN 9781138062177 (hardback : alk. paper)
Subjects: | MESH: Darwin, Charles, 1809-1882. | Mendel, Gregor, 1822-1884. | Galton, Francis, 1822-1911. | Biological Evolution | Genetics--history | History, 19th Century
Classification: LCC QH361 .G35 2018 | NLM QH 361 | DDC 576.8/20922--dc23
LC record available at https://lccn.loc.gov/2017022246

Visit the Taylor & Francis Web site at
http://www.taylorandfrancis.com

and the CRC Press Web site at
http://www.crcpress.com

Dedication

For: Jonah, Evan, Max, Alec, and Arthur

Mendel as an abbot is entitled to design his own coat of arms. The description of these four panels is in the text page 109. Most bishops, including the Pope, have a personal coat of arms. They typically adopt, within the shield, symbols that indicate their interests or past services.

Contents

The main characters

Charles Darwin (1809–1882) The great naturalist who proposed a theory of evolution of species by natural selection.

Gregor Mendel (1822–1884) A Catholic monk and priest who demonstrated a theory of inheritance in the edible pea.

Francis Galton (1822–1911) A young cousin of Darwin by 13 years who tested Darwin's theory of inheritance in rabbits with negative results.

Joseph D. Hooker (1817–1911) A botanist and director of the Royal Botanical Gardens at Kew. One of Darwin's best friends and advisors.

Thomas H. Huxley (1825–1895) A biologist and bulldog defender of Darwin's theory of evolution by natural selection.

Alfred R. Wallace (1823–1913) The codiscoverer of natural selection with Darwin.

George Romanes (1848–1894) Darwin's young student and research assistant.

Hugo de Vries (1848–1935) A botanist who rediscovered Mendel's theory. He developed his own cellular mutation theory to explain sudden changes in plant structure.

Carl E. Correns (1864–1933) A botanist who rediscovered Mendel's theory of inheritance.

Eric v. Tschermak (1871–1962) A botanist who re-discovered Mendel's theory of inheritance.

William Bateson (1861–1926) A biologist and bulldog defender of Mendel's binomial theory of inheritance.

Archibald Garrod (1857–1936) A doctor at St. Bartholomew's Hospital, London who found Mendelian ratios in his study of rare inherited diseases.

Karl Pearson (1857–1936) A student of Galton's, a mathematician and statistician who invented the Chi^2 test among others. He defended

Darwin's blending theory of inheritance during the Mendelian–
Biometrician debate of 1904.

Walter F R Weldon (1860–1906) An Oxford zoologist who believed problems raised by Darwin's theories were purely statistical. A colleague
of Karl Pearson and bitter rival of William Bateson in the Mendelian–
Biometrician debate of 1904.

List of figures

Preface

This book is not written by a professional historian but more from the position of an experimental scientist for whom it is important to formulate the right questions, to design experiments that might be expected to answer them, to do experiments that ask the question of what would happen if..., to build a coherent framework of interrelated ideas, to see how they need to be tested; and then what to do with facts that go against them. This book shows how personal relationships help or hinder progress of such a research project.

More than dealing with the results and explanations of science this book deals with the drama of doing science. This often involves the anguish of going up dead-ends by working on wrong ideas and being unable to drop them when the evidence points in another direction. Or coworkers fall out because of rivalries, jealousies, egocentricity, and thwarted ambitions. Methods are used that are too insensitive to give the correct answer but tantalizingly point to a possible solution.

The text deals with historical characters, so the reader has a right to know which parts are biographical. The majority of events described in this book were taken directly from original sources such as autobiographies, letters, diaries, publications by the actual characters, and eyewitness accounts of public meetings, although not always presented in their strict chronological order. I have used the diary extracts mainly to illustrate the text and have not, therefore, included the dates when they were written.

Similar to many eminent Victorians, Charles Darwin, Thomas Huxley, George Romanes, and Francis Galton were very reticent about their personal lives; for example, it is very difficult to find any personal comments about the quarrel between Darwin and Galton as described in Chapter 6 or the dispute between Romanes and Wallace in Epilogue 2. It is impossible to catch the right tone of people one never knew with only incomplete records such as letters and diaries to go on. The closest intimate records I have used are the diaries kept by Galton's wife, his sister Emma Galton, and Charles Darwin's wife; but these in many ways are more interesting for what is omitted than their actual entries. The diaries mainly consist of brief headings and notes written in shorthand for various topics such as births, marriages, travels, occupations, property, illnesses, and deaths. Darwin, Wallace, and Galton wrote autobiographies, mainly for posterity, and these have been used extensively. The wives of Romanes and Bateson published accounts of their husbands' lives, which have been frequently quoted.

Mendel came from a small farming village in a remote corner of Silesia in Eastern Europe. There are not many accounts of his life there. He took his vows as a monk at the age of 21 years and lived in his monastery until his death in 1884. His social life was considerably restricted. He published at least three scientific papers, known to have written at least ten letters to colleagues (whereas Darwin wrote more than 15,000 letters and Galton more than 450), and Mendel went abroad for the first time at the age of 40 to visit London. I have used his standard biography by Iltis (1924) referenced at the end of this book as well as other named texts there for many of the facts; other details come from the Mendel Museum in Brno. In respect for Mendel's ecclesiastical career as an abbot of St. Thomas, Brno I have selected as many chapter headings relating to religious texts as possible to reflect the topic of each chapter.

The story is mainly told through the eyes of Francis Galton (no relation to the author) because he was involved with the key players throughout his long life. He was an exact contemporary of Gregor Mendel both being born in 1822 and lived until 1911, so witnessing the foundations of modern genetics. Galton came under the influence of his cousin Charles Darwin from the start, which affected his ideas on heredity and my book begins at this stage. Galton and Mendel both presented their first observations on heredity in 1865. Galton's paper was titled *Heredity Talent and Character* published in the *Macmillan's Magazine*, and Mendel's was *Experiments in Plant Hybridization*. Both the titles give a clue to their different approaches to science. The former uses abstract words that are difficult to define such as talent and character, whereas Mendel tells you exactly what he did with plant hybrids–experiments. Comparing Mendel's work with the work of Darwin, Galton, Romanes, and Pearson like this reveals in its true light the genius of Mendel. Mendel unfortunately did not live to see his own work being accepted; he died in 1884. Galton subsequently played a significant role in the progress of genetics and lived to see Mendel's theory firmly established, following the Mendel–Biometrician showdown of 1904 in Cambridge, United Kingdom as described in Chapter 15. He also appears to have recognized how close he had got to Mendel's interpretation with his own studies on plants (the sweet pea) and animals (Basset hounds).

This book would be thus characterized as an account of the investigations by Darwin, Galton, Mendel, Romanes, and others on the nature of heredity in the latter half of the nineteenth century. It is a mixture of the story of science with a good deal of the methods used to obtain the results of science by observation, experiment, and logic.

The epilogues are much longer than usual and I have included other relevant ideas that do not fall easily into the main text by superscript numbers indicating a short note to be found at the end of this book. The sources of the most of the described events are given in a reference list of books and articles I have consulted at the end. There is also a gene timeline of events taking the reader up-to-date with the discoveries of the double helical structure of DNA, cracking the genetic code, and the completion of the human genome project.

My grateful thanks to John Harris, George Davey-Smith, John Walker-Smith, Gordon Ferns, Brian Shine, Tim and Jon Galton, David James, Robert Dudley, and Mary Seed for their patience and good humor for criticizing all or parts of the text. My debt is immeasurable to my teachers J.B.S. Haldane for genetics, J.Z. Young for evolutionary biology, M. Rodbell, Nobel Laureate 1994, for sharing his enthusiasm for scientific research, Russell Fraser for guidance in clinical medicine, and not least to my students without whom very little would have got done.

David J. Galton
Wolfson Institute of Preventive Medicine

Prologue

...bald parents usually have bald children, grey-eyed parents, grey-eyed children and squinting parents have squinting children.

—Airs, Waters, Places. Hippocrates, (c.460–c.377 BC)

We biologists have been from earliest times in great confusion about the mechanisms of inheritance. Our ideas up to the nineteenth century had not progressed much further than Hippocrates as quoted above. The views on heredity in the nineteenth century were still vague and contradictory and the subject had never been squarely faced by scientific experiment. There seemed to be no way of predicting which features of parents (eye color or squints) would be transmitted to their children and in what proportions. In some ways, this was very fortunate. It meant that there was a big problem waiting on our doorstep to be solved. Biologists in the nineteenth century still read Aristotle (384–322 BC) too much and relied on his authority that taught in his book *On the Generation of Animals* that inheritance with two sexes depended on their copulation and the growth of a preformed structure (a tiny homunculus) in the male seed. The male seed did not need to fuse with any female material for its development. The female menstrual blood may be needed to activate the male seed or the female may be just acting as an incubator. To believe in Aristotle is one thing—to grasp a new scientific idea is quite another. So, heredity was an area that many nineteenth century scientists thought important to tackle.

Was heredity an important problem? The biologists who took it up thought so and cared about it more than anything else in the manner of truly committed scientists. If we knew how it worked and when we learnt how to control and manipulate it, we may be able to change the whole future course of human development, altering and improving all the inherited features we pass on to our children. It was one of the burning issues of the time, and still is.

Not to exaggerate, the knowledge of how heredity works and our ability to control it would put man back at the center of his world again. Throughout history mankind is always being displaced from the center of the action. First, we were peripheral to God's unknowable purpose, kneeling in the great medieval Cathedrals to worship his works of creation. In the Renaissance, we recovered our sense of importance, as man became the measure of all things. In the Enlightenment, we learnt to question the authority of Plato, Aristotle, Galen, and others to go to look for ourselves. Then the new cosmology placed us on a wandering planet, one among billions, as tiny specks of life with no more of significance than an atom of cosmic dust. Now, we may be suddenly able to rise from our obscurity and take control of our own destiny.

By the mid-nineteenth century, several dedicated biologists were bent on discovering how heredity works. There was Charles Darwin (1809–1882) and his younger cousin Francis Galton (1822–1911); there was Carl von Nageli (1817–1891), a Swiss botanist who developed some idiosyncratic ideas about a hereditary material that he appropriately called the idioplasm; and there was a very clever German biologist, August Weismann (1834–1914) working in Freiburg. These were all well-educated university men with many students and assistants.

A further aspirant entered the field as a rank outsider, an obscure monk in an Augustinian monastery in the small cloth manufacturing town of Brünn about 120 kilometers north of Vienna. He had no university degrees and had no professional assistants to help him. It seemed no one was aware that he was working on the project until he presented his results in 1865—which everyone ignored. The 150th anniversary of Gregor Mendel's presentation of his theory of inheritance is celebrated in this book. He was unlike the wealthy gentlemen scientists of England and Europe, coming from quite a different social class. His parents were German-speaking peasant farmers at a tiny hamlet called Heinzendorf on the flat Silesian plains of Eastern Europe. The hamlet is about 20 miles from Brünn in the then Austrian–Hapsburg Empire. The parents had three children; a daughter Veronika, a middle son Johann (later changed to Gregor when entering the monastery), and the youngest daughter Theresia. Johann was born in 1822, and so was an exact contemporary of Francis Galton. He went to the local elementary school and because he seemed bright was sent to the local gymnasium—a secondary school to train students for the university. He ran short of money for living expenses around this time but his younger sister helped him out by giving Johann part of her share of the family farm—the part that was to have been her dowry.

If one were to lay bets on who would discover the solution to the heredity problem most people would choose Charles Darwin, an already world famous naturalist who had developed the theory of evolution by means of natural selection. He had the time, patience, and resources to do any of the experiments that the other aspirants could do. He also had a very intelligent cousin, Francis Galton, who volunteered to help him.

My book will mainly focus on the most likely winners—the British scientists—and the most likely loser, the Catholic monk. This book starts with the relationship between Darwin and Galton. They both were searching for an answer to heredity because each had written a celebrated book, *The Origin of Species by Means of Natural Selection* (1859), and *Hereditary Genius* (1869), respectively. These books only made sense if the authors could give a satisfactory account of how bodily characteristics, such as mental abilities, were passed on from the parents to their offspring. This method of transmission of bodily features from generation to generation became the nub of the problem and eventually might show, according to Darwin, how new species might emerge.

Gregor Mendel worked in solitude on the inheritance of plant characteristics (such as seed color, petal color, height of plant) in the edible pea for 8 years without giving any clues as to his motivation. Perhaps he was studying nature (or reading the book of God), for no other worldly reasons than to fulfill his devotional duties because he was a monk (unusually also a priest) in an Augustinian monastery. No one knew he was working on this problem until presenting his results in 1865 to a scientific society in Brünn (now called Brno in the Czech Republic). His findings were published in 1866 and forty reprints were then available for Mendel to publicize his work to the scientific community. He almost certainly sent a copy to Darwin (why would he not?) since he possessed and had read a copy of Darwin's book in German translation on the *Origin of Species* that he had personally underlined in parts with pencil. His reprint showed that he had discovered a mechanism for the inheritance of plant features involving the transmission of discrete, countable particles. Features of the edible pea, such as pea color or shape, that seemed to be lost in one generation cropped up again a generation or two later in their original form that is skipping a generation. He also found a mathematical ratio recurring in his counts of the various inherited features. He had discovered a new constant in the form of a ratio of 1:3 that related to the inherited features of a plant. Discovery of such natural constants is always an important event. What did this ratio mean and where did it come from? What did it say about how these traits were inherited? Mendel created an algebraic model for inheritance that could explain how this ratio of 1:3 came about.

Mendel's particles (or elements) were eventually called genes from 1909 onward. His model turned out to be a correct solution on how heredity can work. By the age of 21 years Mendel had entered the monastery of St. Thomas in Brünn (now called Brno) and taken the vows of chastity, poverty, and obedience. He was ordained as a priest in 1847 at the age of 25. This means that there is very little social life to record, which perhaps allowed him more time to concentrate on his scientific work. The final result of the *race* is announced in the Prologue—so we can stop reading and go home. No. Unlike sporting events, who wins and who loses is not the real issue. Finding a solution to an important scientific problem is never finished. The journey always goes on with fresh questions being raised by the preceding correct answers. And unlike sport, it is always more a cooperative than a competitive effort—even for a solitary scientist as Gregor Mendel. His results had to be confirmed by others before acknowledgment of how truly great his work was. This is what makes science such a fascinating field to pursue.

The definition of the word gene has varied from decade to decade as more information is discovered. Mendel called his discovery *Elemente*, in German, perhaps an element or factor in English. If a gene is defined as a *unit of heredity controlling a particular inherited characteristic of an organism* then Mendel discovered the gene. He had no idea what it was made of (we now know it is DNA), where it was to be found (on chromosomes in the nucleus), or how it worked (by making RNA), but he had built up a theoretical model that revealed some of its essential properties. The most talented scientists of the time in the British team (Darwin, Romanes, Galton, Huxley, Hooker, Pearson, and Weldon) did not appear to understand or act on Mendel's results and this book attempts to explain why. At least the British were one of the first to confirm Mendel's results in humans 34 years after Mendel had published. This gives some idea of the pace at which science often progresses.

Why the British team lost out arose partly from the different theories that Darwin and Mendel had conceived. Darwin and his assistants adopted and worked on a blending theory of inheritance in which multiple uncountable inherited particles, which Darwin called *gemmules*, are shed in varying numbers from every organ in the plant or animal for onward transmission to the progeny through the sex organs. Mendel adopted a binomial-type theory in which discrete particles (which he called Elemente, genes in modern parlance) are transmitted from parents to offspring in binomial proportions. Binomial here means an algebraic expression involving two terms, which Mendel called *Dominant* and *Recessive* (see Chapter 13). Mendel formulated the complex phenomenon of reproductive inheritance in algebra as a simple 1:3 ratio and this was and still is an astonishing feat of science.

Due to Darwin's great reputation and influence on others in the field he persuaded Francis Galton and other colleagues to work on his own theory until his death in 1882. Some of Galton's experiments got very close to Mendels'. However, even in Galton's publications he still published Darwin's blending theory, though he knew it was incorrect—an instance of how the progress of a project can be held up by the hero worship of a famous senior colleague. Unfortunately, Galton and others kept doing the wrong experiments using Darwin's theory until Romanes death in 1894.

There the matter stood until the early 1900s when three European botanists rediscovered that the results of Mendel's earlier experiments were correct. They all found numerical proportions of 1:3 in inherited plant characters after a variety of plant-breeding experiments. Meanwhile, a doctor, Archibald Garrod, at St. Bartholomew's Hospital (where I work) in London also found the same ratios in his studies of the inheritance of a number of rare diseases in first-cousin marriages. The stage was therefore set for a clash in theory between the Mendelians and Darwinians (sometimes called Biometricians) in 1904. The conclusions from all available evidence at the time favored the Mendelian interpretation, although some held to Darwin's blending theory for a few more years.

But to go back to the beginning when Charles Darwin was a 19-year-old medical student at Edinburgh University, when Johann (later changed to Gregor) Mendel was a small boy of 6 years helping his parents on their small farm on the flat plains of Silesia and always short of money, and Francis Galton, also 6-years old, was being spoilt by his wealthy parents in a grand house in the suburbs of Birmingham, United Kingdom.

About the author

David J. Galton is an emeritus professor at London University from the Departments of Molecular Genetics and Metabolism, St. Bartholomew's Hospital. He gained doctorates in medicine (MD for work done at the National Institutes of Health, Bethesda, Maryland) and in science (DSc). He was elected to the Jephcott European Fellowship of the Royal Society of Medicine, London and to the Erasmus European Professorship. He has been the chairman of *Clinical Science*, HEART UK, secretary of the European Atherosclerosis Society, and vice president of the Galton Institute, London, among other administrative posts. He has published 8 books, written more than 400 research publications on genetics of human disease, and served on the editorial boards of the *European Journal of Clinical Investigation, Clinical Science,* and *Metabolism and Cardiovascular Disease.* He is elected Fellows of the Royal College of Physicians and of the Royal College of Ophthalmology London and has served as a consultant physician to St. Bartholomew's Hospital and Moorfields Eye Hospital, London, UK.

chapter one

Seeds of hero worship

A Statue of a Hero with legs of iron, its feet part iron and part clay.

Daniel 2.34

It probably began as one of those intense emotional crushes that young boys sometimes feel for older cousins. It gradually developed into one of the worst cases of hero worship perhaps ever recorded. Even at the age of 64 years the hero worship burned as strongly, and Frank publicly confessed in a lecture that: *I rarely approached his general presence without an almost overwhelming sense of devotion and reverence, valuing his encouragement and approbation more perhaps than the whole world besides. This is the simple outline of my scientific history.*

Frank first met Charles at an impressionable age—Frank was just 6 years old, Charles, a medical student, was 19 when he came to visit the Galton family in Sparkbrook on the outskirts of Birmingham in 1828 (Figure 1.1). The Galton's estate owed their wealth to Frank's grandfather on his father's side who was something of an anomaly. He had amassed a large fortune in the manufacture and sale of arms for the Napoleonic wars of 1808–1814. He also professed to be a good Quaker promoting pacifism and the renunciation of war; he argued that what his customers did with his products was their affair and that guns might even deter conflict. This did not satisfy his colleagues and he was expelled from the Birmingham Society of Friends for *fabricating instruments for the destruction of mankind*. Frank's father, Samuel Tertius Galton, inherited a large part of his grandfather's wealth and had added to it by fulfilling the duties of a competent banker in Birmingham. He had founded a successful bank in Steelhouse Lane that enlarged their fortunes still further. He was happy in marriage to a joyful and unconventional young woman, Violetta Darwin, who was the daughter of Dr. Erasmus Darwin (1731–1802), a talented physician and a published poet of some distinction. Erasmus Darwin was the father of a well-to-do family doctor in Shrewsbury, Dr. Robert Darwin, whose son Charles was the cousin under whose spell Frank fell.

By 1827, the Galtons were living at the Larches, a fine country residence in the Sparkbrook district of suburban Birmingham. The name Larches came from two exceptionally tall larches that guarded either side of their driveway, and Francis (who was always called Frank by his

Figure 1.1 A portrait of Charles Darwin (1809–1882) by George Richmond in the 1830s.

family and close friends) was attracted ever after to this type of tree. The house was a handsome three-storied Georgian building with two ample side-wings, numerous outhouses, and paddocks at the rear end where his numerous brothers and sisters could ride their ponies. One of Frank's first memories in childhood was falling off his pony into a very muddy ditch and being dragged out his feet first by his eldest brother.

In that summer of 1828 when Frank first met Charles the Galton family was already large. Out of the 10 children, 7 had survived into late childhood. Frank had two elder brothers, and then came his four sisters. Frank was the baby of the family and excessively indulged by all his sisters, but especially by the third one, Adele—or Delly as Frank had called her from his early infancy. She greeted Frank's birth as a fairy gift and begged hard to be allowed to consider him as her sole charge. His other sisters petted him as the baby but Adele always had the greatest share of his heart. Frank's chief attractions as a child at that time were an imperfect articulation of English, an earnest desire of having his own way, many cunning tricks, and the source of a great deal of noise.[1] Charles Darwin's visit was really at the instigation of Frank's father who was worried about the educational prospects of his eldest son Erasmus named after his grandfather. Charles at 19 years was already enrolled as a medical student at Edinburgh University and the father's hope was that Charles's visit might inspire Frank's brother Erasmus to take up a similarly serious vocation. In reality, Charles appeared to spend more time with his insects than on his medical studies in Edinburgh.

They all used to take tea in the parlor, a high clean rather empty look-ing room at the back of the house. Frank's mother asked Charles if he wanted to go riding with them on the following day. No—he would really rather go out walking by himself in the fields at the back of their estate. Charles loved riding but he had recently contracted a passion for collect-ing insects, particularly beetles. What workmanship there is in the frame of beetles; such as living watches concealing the thousand springs and cogs of life. He pulled a pillbox out of his jacket pocket to show them and Frank was astonished by the appearance of the insect. It was about the size of a Brazil nut with a shining brown carapace, just like the shell of a conker. It had white wiggly stripes going down to a most ferocious look-ing proboscis. Here were two pincers looking like fret-saw blades facing each other, and two long antennae extended similar to curved pylons from the base. Woe betides any unsuspecting smaller insect that acci-dentally strayed between the blades of this ferocious beetle; they would be instantly mashed into little pieces. Charles referred to the beetle by a villainous sounding Latin name that Frank did not understand.

Frank was overwhelmed after tea when he followed Charles, at a distance, into the rear of their estate by the river.

It was a pleasant countryside at that time with gentle hills, the woods were full of fine timber, and the valleys beyond were comfortable and snug with rich meadows, and several neat farmhouses scattered here and there. Sadly the town of Birmingham has now encroached on all this farmland. Charles was scraping at the bark of a rotting old tree by the riverbank. Two rather dull colored beetles scampered out from a crevice and were immediately captured by Charles. One fell to the ground onto its back and its legs struggled similar to an orchestra playing Beethoven. The other, a gigantic black beetle tried to escape and Frank was thrilled to see Charles place the smaller of the beetles into his mouth to free his hand to capture the giant. A few seconds later, he spat it out. He told Frank and the others afterward that the beetle had spurted out an acrid juice from one of its body glands that tasted foul.

Charles loved collecting beetles and other insects. He might even have died for his love of them. While in South America exploring the province of Mendoza on the Beagle expedition he was attacked by the aggressive black assassin bug. This is a species of reduviid insect, the vinchuca bug that lives in the roofing and thatch of local houses. It was the most disgusting thing to feel the soft wingless insect about an inch long drop down from the roof and crawl over your body at night, quite fearlessly darting at any exposed skin surface to suck blood. Charles rather foolishly caught one the next day and placed it on a table, and although surrounded by people, the bold insect charged at his bare fin-ger brandishing its sucker to draw blood. The wound caused no pain and it was curious to watch the insect's body during the attack change from wafer thin to a globular one bloated with blood. This may be linked to Charles developing a long chronic illness in midlife that caused palpitations, shortness of breath, and he intermittently suffered stomach disorders for the rest of his life. At the age of 33 years, he had gone on a long tour of North Wales to study the geological effects of the extinct glaciers that formerly filled all the larger valleys. This was the last time that Charles ever felt strong enough to climb mountains or to take such long walks that these expeditions required, due to his shortness of breath and giddiness. No proper diagnosis was ever made. Charles went for various water cures with variable results, but the general conclusion was that his symptoms were due to some form of hypochondria.

The doctors might have misdiagnosed him, because by the beginning of the twentieth century, it was found that the reduviid bug transmits an infectious parasite that lodges in the heart and the lining of the intestines. The main features of this parasitic infection, Chagas disease, fit like a hand-in-glove to all the symptoms that Charles developed. A patient with Chagas disease often has a dilated heart with failure of the circulation producing breathlessness and fatigue; the intestines are affected leading to severe indigestion and abdominal distension; and to this day there is still no completely satisfactory treatment.

Of course, Frank copied Charles in his passion for collecting insects, but extended his collection to seashells, minerals, and coins. The beetles were his treasured possessions. A last will and testament was found in an old trunk that Frank wrote at the "advanced" age of 8 years, bequeathing his insect collection to his dearest sister Adele. He left his mineral and seashells to another sister Bessy.[2]

In the following days of Charles's visit, another incident occurred that had a big influence on Frank. His father had a scientific bent and as a banker had published a general paper on money supply, price level, and the exchange rate but without really clarifying the relationships that were involved. His great respect for science led him to collect all sorts of scientific instruments, although he probably could not tell you the difference between a theory, a hypothesis, or a concept and had no idea about the basic principles of scientific method. He collected objects such as antique telescopes, armillary spheres, and astrolabes, which were scattered around the house on shelves, taking the place of the usual domestic ornaments such as vases and porcelain figures. One highly prized instrument was an intricate eighteenth century vernier barometer housed in a beautiful inlaid wooden case. Charles wanted to examine it and Frank's father took it off the wall and very patiently tried to explain how it worked. Frank was hanging around in the background keeping close to his older cousin. Frank did not understand how atmospheric air could weigh anything or how it could depress a column of mercury. The brass vernier scale was also beyond him. But the whole episode awakened a sense of mystery and delight in strange and exotic scientific instruments that remained with Frank ever after. As an adult Frank invented some instruments of his own and he tried to live his life by the scientific method. All aspects of his life were to be treated in the spirit of an experiment and to be measured, if possible. It provided a certain measure of detachment in his personal relationships. For example, if he approached an attractive woman at a social gathering, the encounter was treated as an experiment. How would it turn out? He remained as objective as any field observer and had no particular desire to see one outcome prevail over any other. He would just vary the conditions of approach at the next encounter to see if it would turn out differently. In the end, he invented an instrument to record the sexual attractiveness of all the women that he passed in the streets. He kept the instrument in his pocket and rated the women as they went by. He constructed a "beauty map" of the British Isles and found the least attractive women to be in Aberdeen; the most attractive were to be found in London.[3]

From Frank's father's point of view, Charles's visit was not particularly successful. Frank's brother Erasmus had nothing much in common with Charles. Indeed, they hardly spoke to each other during the whole stay. Erasmus was determined anyway to go into the Navy. However, the visit did succeed in influencing Frank; he wanted to copy what his cousin Charles was to do. Frank's mother was also keen for her son to study medicine because her father had been a very successful medical practitioner and she hoped to see the profession carried on in the family. So, from an early age it was always to be medical studies for Frank too.

Mendel's childhood was mainly spent on his father's farm where he helped to tend the orchards and became very interested in bee keeping, to make honey. His interest in bees survived to his time in the monastery where he looked after batches of hives to produce honey for the brethren. In childhood there is no record of him having had any scientific mentors.

To be a doctor[4]

It was Darwin's father who drove Charles to study medicine. *You care for nothing but shooting, dogs and rat-catching; and you will be a disgrace to yourself and all your family*, Charles was once told in a fit of irritation by his father. To avert this dire prognosis Charles was duly enrolled as a medical student at Edinburgh University. He only stayed in the course for 2 years. The subject was of intrinsic interest, but he found most of the lectures intolerably boring. There were long stupid lectures by a Professor Duncan on materia medica starting every morning at eight. They reminded Charles of the method used to detect excessive fluid that can accumulate in the abdomen by percussion of the stomach wall. The procedure is called listening for "shifting dullness," which perfectly fitted the contents of the professor's lectures. Then a Professor Munro lectured on Anatomy. Charles disliked him and his lectures so much that he could not speak with decency about them. The Professor was dirty and slovenly in person as well as in his behavior. It was not uncommon for him to enter the lecture theatre bedaubed with blood from his recent dissections and his teaching was very out-of-date.

The final straw came when Charles had to attend an operation on a small child. The little girl had fallen under the wheel of a cart on Princes Street and she had crushed her right foot. The foot had become infected and then gangrene had set in and now needed to be amputated. There was no anesthesia in those days. The child was just wrapped in a blanket to stop her struggling with her foot protruding at the lower end. She was laid on a bare operating table with small wooden tables at its side on which was assortments of surgical instruments neatly laid out in rows. They resembled the sort of tools one might find in a carpenter's shop: large metal mallets, pincers, strong scissors, and ferocious looking handsaws. One thick metallic saw had deep notches along its cutting edge to trap any bone or flesh from clogging the blade as it sawed through the leg. The surgeon, with his assistants dressed in loose fitting white tunics with rolled up sleeves held the child down, while the chief surgeon commenced the amputation. The eager faces of the students, including Charles, were ranged at the back of the room. Speed was of the essence; if the whole operation could be completed in 10–20 seconds she would have a good chance of survival. The death rate from amputations at this period was about 50 percent. The surgeon started to see into the leg about four inches below her right knee. The crescendo of screams of the little girl was unbearable, eventually subsiding into deep sobs of despair as the child became exhausted. She never lost consciousness throughout the whole operation of about four minutes, but toward the end her cries seemed to be disconnected from the activity of the surgeon—God alone knows what she was really suffering. This horrible and cruel experience was enough to drive Charles completely away from any more medical studies. The cruelty was unnecessary. His cousin, Frank, as a medical student also saw an emergency amputation of both legs of a powerful drayman who had fallen under the wheels of a stagecoach. The man was virtually dead drunk when he was brought into the operating theatre and the amputations were started immediately. He felt nothing, and indeed was in a drunken sleep throughout the whole procedure. One wonders why they could not make all preoperative patients "dead" drunk, so they were spared the pain of the surgery. Anesthesia was a marvelous invention but did not come into standard practice until the early 1840s with laughing gas (nitrous oxide), then in the mid-1840s with ether, and eventually in 1847 with chloroform.

Charles spent more and more time in Edinburgh studying his beloved insects and made some very interesting and original observations on the habits of marine invertebrates. He joined the local Plinian society to present some of his findings. He met there Robert Edmond Grant (1793–1874) who was an Edinburgh-trained physician. Grant had given up medicine to study the evolution of invertebrates and had even cited Erasmus Darwin's *Zoonomia* in his medical dissertation. During the Plinian Society's joint collecting trips to the sea shore, the older Grant introduced Darwin to the world of research and microscopic dissection—and this led to the Darwin's first scientific paper, delivered at the Plinian Society in 1827.

As Darwin later wrote in his autobiography, *He* [Grant] *one day, when we were walking together, burst forth in high admiration of Lamarck and his views on evolution. I listened in quiet astonishment, and as far as I can judge, without any effect on my mind.* Later in life, Darwin would distance himself from Grant, probably because of Grant's radical views on the transformation of species following on from Lamarck's ideas on the inheritance of acquired characters. Lamarck believed that if, for example, a father developed his musculature during his work as a blacksmith then his children would inherit as strong a musculature from him. Incidentally, Hippocrates antedated Lamarck's views on the inheritance of acquired characteristics by writing *characteristics thus acquired* (referring to the custom of molding the heads of newborn infants to an elongated from the spherical form) *at first by artificial means, but as time passes becomes an inherited characteristic so the practice...* [of binding the head]*... was no longer necessary.*

Darwin dropped out of medical studies altogether by 1828.

Frank's experiences studying medicine were quite different. He started at the Birmingham General Hospital in 1838 and rather enjoyed the charade of medicine. He particularly liked working in the dispensary. He became adept at making a variety of infusions, decoctions, and extracts of various herbs and minerals. He was never quite sure what good they did, so he started to try them on himself. He developed quite a taste for one particular decoction: a quart of *aqua vitae*, one ounce of aniseed bruised, one ounce of liquorice sliced, and half a pound of poppy seeds (from *Papaver somniferum*) all steeped for 10 days, after which the supernatant is poured off into a bottle containing two tablespoons of fine white sugar. It was meant to be a cure for asthma, but he found it quite a decent cordial after evening meals. He did worry afterward about how much opium was contained in the poppy seeds. He never felt any drowsy effects. He tried many of the drugs in the materia medica but croton oil stopped him dead in his tracks. Explosive vomiting and a cataract of watery diarrhea completely cured him of trying out any more of the other pharmaceutical preparations. More serious duties were gradually imposed on him. He had to go with the surgeons on their morning ward rounds and then to attend the accident room. On the ward rounds he made notes of the cases and he wrote the prescribed treatment as dictated by the qualified doctors on a sheet of paper fixed to the head of the bed. He appreciated from the very first the importance of a careful study and record of every patient. He remembered one episode very clearly. A young girl was dying from typhus and he had been instructed to apply a mustard plaster. When he came to her she was fully conscious and she said in a faint but perfectly composed way: *Please leave me in peace. I know I am dying and am not suffering. There is nothing you can do for me.* He knew she was right and had not the heart to distress her any further.

In the accident room, his main duties were bandaging and plastering the victims of various accidents. Lacerations of the arms or legs needed tight bandaging to staunch the blood loss. Torn scalps from brawls were quite common at the weekends and his job was to shave the head and then to stitch the wounds together with a three-cornered "glove needle" that cuts its way through the skin. He also did cuppings and became quite good at tooth-drawing. He set broken limbs and occasionally had to reduce dislocations of the shoulder joints.

Why Frank thought medical practice contained a good deal of charlatanism came from his experiences with a Dr. M. who boasted of having no scientific acquirements and knowing next to nothing of anatomy or physiology. He always got his patients out of the hospital more quickly than his colleagues. His treatment was simple and invariable. It consisted of a strong purgative followed by a starvation diet and then subsequent overfeeding as soon as all fever had gone. The composition of his "drench" never varied whatever the illness he was treating. A big bottle was made of it every morning in the dispensary; it was so cheap that any surplus could be thrown away and a fresh infusion made for the next day.

After 2 years in Birmingham, Frank's father sent him to continue his studies at King's College London. Frank went to stay with four other students in the house of Professor Partridge. The professor lived near Charing Cross and was a brilliant man of about the age of 34. He was currently engaged in bringing out an encyclopedia of physiology, which was a remarkable project in those days. Frank enjoyed his studies with him and found the whole level of teaching at Kings College to be far superior to that of Birmingham. However, the chief attraction for him to come to London was that Charles had returned from his travels around South America and was living not far away in Upper Gower Street. He had worked up his travel journals for publication that eventually became known as *The Voyage of the Beagle*, and Frank frequently went to see the fossil animals that Charles had collected as well as to hear all about his journeys and adventures in South America. It was all too natural that Charles's passion for travel captivated Frank with the force of a migratory bird.

And he was able almost immediately to indulge this passion, as his father died prematurely leaving him with independent financial means so that he no longer needed to pursue his career in medicine.

Meanwhile, Mendel had completed his primary education at his local village school in Heinzendorf. Then his parents decided to send him to a nearby town (Troppau) to complete his secondary education at the local gymnasium—a type of secondary school to train students for the university. His father expected that Gregor would run the family farm and the education would prove invaluable for the rapidly changing agricultural practices of the time to improve crops and animals.

He graduated from the Gymnasium aged 19 with sufficient honors to gain entry to the Philosophical Institute in the nearby city of Olmutz to do the 2-year course required of gymnasium graduates before they could begin their university studies. This was a hard time for Mendel; he became homesick and depressed and was continually short of money for living expenses. He undertook private tutoring and his younger sister Theresia generously gave him a part of her share of the family farm—the part that was to have been her marriage settlement. This did not prevent her marrying a Jacob Schindler and she had three sons by him, Alois, Ferdinand, and Johann. Mendel had a life-long gratitude to his sister for her help and took a great interest in supporting his three nephews when they all lived in Brünn. Even so Mendel still could not make ends meet. At the suggestion of his physics teacher at the institute, Professor Friedrich Franz and Mendel, in common with many other penniless young men in search of an education, was advised to enter a monastery and to become a monk. Professor Franz, who was a priest himself, advised Mendel to go to the Augustinian monastery of St. Thomas in the city of Brünn. This was a wealthy industrial city about 120 km from Vienna mainly involved in the manufacture of clothing and other textiles. It was nicknamed the "Manchester" of the Austrian Hapsburg Empire. The population was about 70,000 and similar to some other cities in Austria there was a conflict between a Czech majority wanting to maintain their language and culture with a German minority in power prohibiting the use of Czech in many of the good schools. Brünn had an orchestra, a philosophical institute, and a new technical university. When Mendel arrived there was also an agricultural society founded by a group of amateur naturalists in the early 1800s and since 1827 its president was Cyrill Napp, Abbot of the Monastery in Brünn.

To be an explorer[5]

Charles Darwin made far better use of his travels than Francis Galton ever did. Darwin's experience provided the inspiration for all his later work; it was the formative period of his life. Galton explored the darker regions of Africa more for the love of adventure; he naively wanted to be an explorer similar to David Livingstone, and had pleasurable expectations of hunting big game. Galton gathered a few travelers' tales to tell about his explorations, whereas Darwin developed powerful and broad-ranging ideas from his expedition that challenged and changed the intellectual climate of England and the rest of the world.

Some of Darwin's ideas probably came from his grandfather, Erasmus Darwin, who had published a book in 1794 called *Zoonomia* or *the Laws of Organic Life*. In it, he stated his belief in the transformation of species from one type to another, and he believed that modification of species was brought about by the satisfaction of their internal drives as they adapted to their environmental conditions. Lamarck (1744–1829), the great French biologist, had published *Philosophie Zoologique* (1809) in which he stated that species grade into one another by a process of evolution (or transformation as he called it). Due to this modification of species during long periods of time the whole of the animal kingdom could be represented by a genealogy of branching lines such as a tree, the last branch being that of a man. He thought that some fossil animals had not become extinct but had been transformed into their now living descendents. Many other lesser known biologists had adopted and published their views on the evolution of species before the Beagle sets sail (Geoffroy Saint-Hilaire, W. Herbert, Professor Grant, and Herbert Spencer to name just a few); so animal evolution was very much "in the air." What Charles Darwin managed to find out was a plausible mechanism of how evolution might work by the process of natural selection.

As for exploration, Darwin had been fired by the popular travel books of Alexander von Humboldt (1769–1859). Von Humboldt was an explorer and naturalist, traveling widely in Central and South America between the years 1799 to 1804. He discovered a vast number of new plants and collected all the information on their habitats. He spent the years from 1804 to 1827 in Paris writing up the account of his expeditions. He was a pioneer in trying to relate the geography, geology, and climate of a locality to the plants and animals living there. He was among the first ecologists and Darwin was to develop a life-long interest in this subject.

After leaving Edinburgh failing to complete his medical studies Darwin was sent to Cambridge University with the intention of becoming a clergyman. There Darwin was invited to travel as an unpaid naturalist and geologist aboard a government ship, the Beagle, and was sent to explore and make a survey of the coast of South America. His father was dead set against the voyage but, with the help of his uncle Jos Wedgwood, Darwin managed to win his father's approval. They set sail from Plymouth on December 27, 1831 when Darwin was 22 (and Galton was 9 years old).

Her Majesty's ship, the Beagle, was a ten-gun brig under the command of Captain FitzRoy. The captain was a methodical and serious man taking 24 very good chronometers on board to calculate the exact longitude of Rio de Janeiro that was still unknown, and to chart the coastal waters around the rest of South America. The Panama Canal was still a plan on paper at the time and many shipwrecks occurred on the sea routes around Cape Horn. Darwin's job was more to chart nature's handiworks and he began by collecting specimens of the fauna and flora from each port of call. He collected an enormous number of animals, plants, and fossils. In Rio he was "red hot" for spiders; in Punta Alta for fossil bones; in Galapagos for finches and other birds.

He collected insects, especially because of his love for beetles, and also reptiles, marine animals, seashells, minerals, and plants from everywhere. He was indefatigable. The collections were all methodically labeled and sent back to England on homeward bound vessels. However, these were not mindless collections of samples by an amateur naturalist. Facts are not science—just as the dictionary is not literature. Darwin thought deeply and continuously about his collections, and over the years he developed two really revolutionary ideas—evolution by natural selection and the descent of man from animal progenitors.

While visiting the Galapagos Islands for 5 weeks in 1835, one problem forcibly struck him—that is the *mystery of mysteries, the first appearance of new beings on Earth*. Galapagos are a group of more than 19 separate volcanic islands situated astride the equator about 1000 km from the west coast of Ecuador. They had erupted from the sea bed in relatively recent times (that is about 5–9 million years ago); the formation of the main continental land masses of North and South America are about 500 million years old. When Galapagos emerged from the sea they had absolutely no life on them. How did they acquire the diversity of life that Darwin now encountered on the different islands? Perhaps, to start with, seeds drifted there on seaborne or windborne currents from South America and lodged on the islands to provide meager vegetation. Perhaps sea birds blown off course by gales arrived, settled, and started to spread more of the plant seeds after eating the fruits to other parts of the islands. The vegetation flourished on some of them and the original founder birds multiplied and diversified. Darwin observed that the birds always bore a striking resemblance to their nearest relatives on mainland America. There was no evidence for independent creation of a completely new species. However, there were some marked differences between the Galapagean birds and their closest relatives (or supposed ancestors) from mainland America. For example, the beaks of the finches on the separate islands were so different from each other that the birds seem to have evolved into distinct species. An extract from his travel journals explains this:

> *Galapagos, The Voyage of the Beagle. Sept. 1835.*
> *The remaining Island birds form a most singular group of finches, related to each other in the structure of their beaks, short tails, form of body and plumage: there are thirteen species, which Mr. Gould has divided into four sub-groups....the most curious fact is the perfect gradation in the size of their beaks in the different species, from one as large as a hawfinch to that of a chaffinch and even to that of a warbler. Seeing this gradation and diversity of structure in one small intimately related group of birds one might really fancy that from an original paucity of birds in this archipelago one species had been taken and modified for different ends.*

The finch population on the islands did not appear to be a stable species. The advantages of the variation of beak sizes may not be immediately apparent, but it can give the individual bird a better chance of exploiting a different food supply. For example, a stronger beak will crack harder and larger seeds. The advantageous trait is passed on to their offspring and this can lead gradually to the formation of a new species, especially in isolated geographical areas such as the Galapagos where there are no mammalian predators (cats, dogs, stoats) to keep the original bird numbers down. Likewise a finer, thinner beak would be more suitable for capturing small insects crawling around the abundant cacti of the islands. Another finch had evolved woodpecker-like behavior and had flourished. No native woodpeckers had arrived from the American mainland in the past, so that this particular ecological niche was not filled. The finch was free to develop into a "woodpecker finch" and adapt to a food supply of insects and grubs found in tree barks. The bird had even learned to use a small twig or cactus needle in its beak to impale and extract small grubs from their holes in trees. Darwin concluded that possibly one species of finch had initially arrived on the islands and then had evolved into multiple new species. This idea was of course heretical. The account in genesis revealed that God individually created all the different species found on our planet within a week.

Darwin's second idea was even more heretical concerning his suspicions about the origin of humans. While visiting Tierra del Fuego, he was astonished to see men and women living in such an abject and savage state. They certainly did not appear to be made in God's image.

> *Tierra del Fuego, Voyage of the Beagle Dec 1832.*
> *These poor wretches were stunted in their growth, their hideous faces bedaubed with white paint, their skins filthy and greasy, their hair entangled, their voices discordant and their gestures violent. Viewing such men one can hardly make oneself believe that they are fellow-creatures and inhabitants of the same world.... I shall never forget how wild and savage one group appeared: suddenly four or five men came to the edge of an over-hanging cliff; they were absolutely naked and their long hair streamed about their faces; they held rugged staffs in their hands and, springing from the ground they waved their arms round their heads and sent forth the most hideous yells.*

Later, Darwin wrote with some persuasion that such savages might indeed be our progenitors; that we had evolved from species similar to them after much social development and modification, rather than having descended from Adam and Eve out of the Garden of Eden.

Galton admired Darwin's travel journal immensely. The intellectual energy of the man and his passion for collecting left Galton completely enthralled. Darwin's revolutionary theories were expressed with meticulous care and close reasoning in his two most influential books on the *Origin of Species by means of Natural Selection* (1859) and the *Descent of Man* (1871). His travel book, the *Voyage of the Beagle* (1839), was a best seller and ran to 20 editions. It has remained in print to this day.

Galton also published an account of his travels in South Africa but it is almost embarrassing to write about them in the same manner as Darwin's. The only ideas stimulating Galton were the thought of hunting big game, exploring, and opening up new tracts of the dark continent. In the 1850s, there were vast blank spaces on the map of Africa and his goal was to explore the country between the West Coast and a newly discovered lake, Lake Ngami, in what is now Namibia.

Galton left England on April 5, 1850 in an old teak-built East Indiaman called the Dalhousie. She was very slow and quite incapable of beating into a head wind. It took them nearly 80 days to reach Cape Town. Galton was no naturalist, but he was lucky to secure the services of Charles Andersson, a young Swede, who spoke English fluently and he became Galton's traveling companion. Galton's published account of the expedition is full of lively anecdotes with a smattering of anthropology. Galton was more interested in the social habits of the tribal people he encountered than the fauna and flora of the land.

The Ovampos tribes were the most interesting of the several groups of the inhabitants that they encountered. Unlike the neighboring tribe, the Damaras, they were kept under very strict discipline by the chief. Galton was not free to do what he liked in their company but had to depend on their wishes. Chief Nangoro was supreme. Galton could not enter the territory, or trade in it, or leave without the chief's permission. Galton tried very hard to make himself agreeable to Chief Nangoro. Before leaving London he had purchased a quantity of beads, trinkets, and other ornaments as gifts of passage for such tribes. In Drury Lane, he had bought a magnificent tinsel crown, a theatrical prop, made out of strong cardboard. On first being introduced to the chief, Galton gravely offered him the crown. The chief bowed his head with dignity on which Galton placed the crown. His head was rather like a bullock's, so Galton patted the crown down with great solemnity to make it sit tight. He looked every inch a king. The chief's entourage went into cheers of delight and Nangoro himself gave every sign of self-satisfaction after seeing his reflection in the mirror that Galton carried with him. However, Galton had to pay for this elaborate ceremony. On returning to his tent in the evening he found a fine-looking buxom negress called Chipanga wearing a great quantity of beads and rings but very little else cavorting around his tent humming sentimental airs to herself. She had a decidedly nice-looking face, very open and merry, but with rather coarse features. Galton was expected to take her as a temporary wife. Her body was covered with red ochre and smeared with butter fat and as capable of leaving a mark on anything she touched as a well-inked printer's roller. Galton was dressed in his last well-preserved safari suit made of white linen, so he had her removed from the tent with scant ceremony. He wrote in his travel book that:

> *This is one of the drawbacks of becoming too friendly with*
> *an African Chief. They expect you to receive a spare wife*
> *or niece in marriage and take great umbrage if you refuse.*

The Ovampos were mainly agricultural people who sometimes employed (or enslaved) Hottentots to help with their animal husbandry. One day Galton encountered a perfect Venus among Hottentot women, and was aghast at the enormous size of her bottom. He had never seen a back-end project so far out. He always professed to be a scientific man and was exceedingly anxious to obtain accurate measurements of her backside; but there was difficulty in doing this. He spoke no word of Hottentot and could never explain to the lady what the object of his foot rule could be if laid across her bottom. He therefore felt in a dilemma as he gazed at her behind, a gift of bounteous nature to this native tribe, which no Mantua worker with all her crinoline and stuffing can do otherwise than humbly imitate. The object of his admiration stood under a tree and was turning her bottom about to all points of the compass as ladies who wish to be admired often do. Of a sudden his eyes fell on his sextant; a bright thought struck him and he took a series of observations on her figure in every direction, up and down, crossways, diagonally, and so forth, and he registered them carefully on an outline drawing for fear of any mistakes. This being done he boldly pulled out his measuring tape and recorded the distance from where he was to the place where she stood. Having thus obtained both bases and angles it was a simple matter to work out the degree of the callipygian (what a beautiful bottom) by trigonometry and logarithms.

Trivialities such as these show the difference between the travels of Darwin and Galton. Galton's book was not a best seller but did run to a third edition. Extracts from a letter of Darwin summed up what he thought of Galton's exploits.

> *Sea houses, Eastbourne, Sussex[6] July 1853.*
>
> *…Last night I finished your volume with such lively interest that I cannot resist the temptation of expressing my admiration at your expedition and at the capital account you have published of it. What labours and dangers you have gone through; I can hardly fancy how you can have survived them, for you did not formerly look very strong, but you must be as tough as one of your own African wagons!*
>
> *I live at a village called Downe near Farnborough in Kent and employ myself in Zoology, but the objects of my study are very small fry and to a man accustomed to rhinoceroses and lions, would appear infinitely insignificant.*
>
> *I should very like to hear something about your brothers: I very distinctly remember the pleasant visit at the Larches and having many rides with them on ponies without stirrups.*
>
> *We have come to Eastbourne for a few weeks for sea bathing with all our children, now numbering seven.*

This letter brings up the topic of their family lives, which turned out to be very different for each of them. Darwin did most of his best work in his family home surrounded by his wife and numerous children. Galton's marriage was childless and he seemed to have spent as much time away from home as possible attending scientific meetings and traveling the world.

Mendel had no such advantages of exploring the world and learning from it. His travels were mainly around the local towns of Silesia, a rather barren corner of the Austrian–Hungarian Empire. He first went abroad at the age of 40 years to visit and to help set up a display stand on crystallization at a technological exhibition in London in 1862.

To be a husband[7]

The marriages of both Darwin and Galton started as very cold-blooded affairs. None of the first fine careless raptures of eloping to Italy, similar to their contemporaries Mr. Robert Browning and Miss Elizabeth Barrett; or dashing off to live in foreign lands in a *ménage a quatre* similar to Byron and Shelley with Mary Godwin and her half sister Claire Clairmont. Darwin's approach to marriage was distinctly cautious. In 1838, he made a checklist of all the pros and cons of getting married in the first place. They ran as follows:

> *To marry*: Children—(if it please God)—constant companion (and friend in old age)—object to be beloved and played with—better than a dog anyhow—home and someone to take care of house—charms of music and female chitchat—These things are good for one's health—but terrible loss of time. Anxiety and responsibility—less money for books and so on. If many children forced to gain one's bread (But then it is very bad for one's health to work too much).
>
> *Not to marry*: Freedom to go where one likes—choice of society and little of it—conversation of clever men at clubs—Not forced to visit relatives and to bend in every trifle—perhaps quarrelling—Perhaps wife would not like London; then the sentence is banishment and degradation into an indolent, idle fool.

The balance clearly came down in favor of marriage. But to whom and when? Preferably a woman who was like an *angel and had money*—like for instance, his cousin Emma Wedgwood. She came from a very good Staffordshire family being the granddaughter of old Josiah Wedgwood, the founder of the Wedgwood pottery factory. He was also Darwin's grandfather on his mother's side, so Charles and Emma were first cousins (see later for Darwin's views on marriage of first cousins and inherited illness). Emma came from an enormous family, which was rather off-putting to Darwin who was less sociable and would have preferred fewer relatives to have to visit and deal with. She was the youngest of eight children and had as many aunts and uncles on her father's side of the family; on her mother's side she had eight aunts and two uncles. Darwin rode over to visit Emma at Maer village in Staffordshire where she was now living, with the aim of proposing marriage. He was not too confident a suitor, thinking himself so repellently plain. It is true that he was not very handsome and his manners required intimacy to make them pleasing. He was too diffident to do justice to himself; but when his natural shyness was overcome, his behavior gave every indication of an open nature. He naturally concealed his former view that he thought a wife would provide a better companionship than a dog. He was anxious and awoke on the night before the proposal with feelings of panic, sweatiness, and a troubled beating of the heart. Next day he plucked up enough courage to propose to her in the library at Maer Hall, their country estate, and was more than surprised to be immediately accepted. He knew she had had at least four previous proposals of marriages and had summarily rejected them all. She was now 31 and many of her friends and relations were wondering whether she was destined for a cheerful spinsterhood surrounded by her numerous family. However, Emma was taken with Darwin; he had some characteristics that Emma really prized:

> *He is the most open transparent man I ever saw and every word expresses his real thoughts. He is particularly affectionate and very nice to his father and sisters and perfectly sweet tempered; and possesses some minor qualities that add particularly to one's happiness, such as not being too fastidious and being humane to animals.*

They were married on January 20, 1839 at St. Peters church at Maer on her father's estate. The bride was given the expected bond of £5,000 and an allowance of £400 a year as long as her father's income could supply it. Of course, this technically became the immediate property of Darwin, because women's property rights were not recognized by parliament until 1882 (The Married Women's Property Act). They appeared to have enjoyed an exceptionally happy and devoted marriage and were blessed with the arrival of ten children.

Galton was equally cautious in the choice of his wife. He protested against the idea that marriage is solely a union of two individuals who are strongly attracted to each other or even to being in love. For him it was more important to marry into a family that was good in character, in health, and ability. Wealth was not an issue with him because he inherited a fortune from his father. For him marriage was more to be considered an alliance of families rather than just of two young people.

Galton first met Louisa Butler in 1853 at a Twelfth Night party in Dover. She was part of an academically distinguished family and the eldest daughter of the Rev. George Butler.. Louisa's father had been a senior wrangler at the Cambridge University meaning that he had a top mathematics undergraduate degree, a position once regarded as the greatest intellectual achievement attainable in Britain. He was then Head Master of Harrow School, before taking his present position as Dean of Peterborough. Her youngest brother, Montagu, became a senior classics scholar at Cambridge University, and was to become Dean of Gloucester and then Master of Trinity College, Cambridge. Her three other brothers all took first class degrees: two becoming head masters of public schools and the other a barrister. Galton visited the Butler House on four occasions, walked out with Louisa several times, and a month later wrote formally to Louisa's father to request her hand in marriage. Galton heard in the affirmative the following week.

Louisa was a plain-looking girl at that time and Galton confessed that his attachment to her was neither romantic nor physical. He was, however, sexually attracted to women and did not like to recall the infection and fever he picked up while traveling in the Lebanon at the age of 23 after a one night of pleasure. His friend Montagu Burton who had just visited Damascus commiserated with him:

> *What an unfortunate fellow you are to get laid up in such a*
> *serious manner for, as you say, a few moments amusement.*[8]

One must guess what he had done.

Louisa was intellectually gifted and genuinely supported with interest on all his projects. She was a great help to him in many ways, but one deep disappointment was that she bore him no children. He probably blamed her for this, conveniently forgetting that his previous infection in the Middle East might have played a part. He came from a large family in which there was always the hustle and bustle of noisy children going about their play and business. Furthermore, Galton admitted to a perfectly detestable feeling that frequently came over him when he read any of Darwin's letters. Darwin never failed to mention that his eighth, ninth, or tenth child had just arrived and that his wife Emma was doing as well as could be expected; but he was still worried about the health of little Etty or Lizzie. It seemed so unjust that Darwin should be able to have so many children while Galton's marriage remained barren.

He may have felt that he had to hold his own in Darwin's esteem, which mattered to him very much. After all, the hero worshipper likes his hero to admire him back. Perhaps he could make his mark by writing as many books about science as his cousin appeared to be doing.

As for Mendel aged 21, he had now joined the Augustinian order of monks at the St. Thomas Monastery in Brünn where he had vowed celibacy. He later became a priest. This was against his father's wishes. Being an only son his father Anton Mendel expected the boy to take over the running of the family farm in due course. "He is a grave disappointment to me" the father would have the cause to exclaim. The son's guilt about this was somewhat alleviated when his eldest sister Veronika married Alois Sturm who agreed to take on the responsibility for the farm.

chapter two

A tale of two books

By their fruit you will recognize them

Matthew 7:16

A scientist's path to fame and glory is by the originality and importance of books and the articles they publish. In the nineteenth century, it was not so much "publish or perish" as it is now to retain one's research post at the university. However, in the nineteenth century, books and research articles were the main ladder to enter the learned societies to ascend to high social position. In many ways the learned societies in the nineteenth century were as important as the universities, and belonging to them gave one as much prestige as being appointed to a top university position. Even now being a Fellow of the Royal Society is held in higher esteem than some top positions at other universities.

Darwin's book—The Origin of Species by Means of Natural Selection[9]

The publication in 1859 of Darwin's *Origin of Species* marked an epoch in Galton's development. It changed his feelings for Darwin from one of boyish hero worship to a state of near reverence.

Although Darwin declined to discuss man's origin in his book: *I think I shall avoid the whole subject as it is so surrounded with prejudice; though I fully admit that it is the highest and most interesting problem for a naturalist,* he implied everywhere that his theory of evolution applied to man too. In fact, he was already compiling a mass of data to write another book on *The Descent of Man,* and *Selection in Relation to Sex.* At one stroke, it seemed he had demolished a system of dogma erected by the theologians and aroused a spirit of rebellion against all the ancient authorities whose positive but unauthenticated statements were contradicted by modern science. This was one of the first times that the authority of the Bible had been challenged in principle about the origins of man.[10] Other stories in the Bible causing dispute could still be resolved by assuming an all-powerful God working miracles. In Joshua Chapter 10, verse 13 it is written that: *the sun stood still and the moon halted... And the sun stayed in mid-heaven for almost a whole day in Gideon.* God could have worked miracles to protect the Earth from massive tides and overheating of the Earth's side that faced the sun. However, Darwin's new theory on the descent of man directly contradicted the account of creation in genesis. It appeared that man was not created in God's image but was ascended by gradual modification from a progenitor who was a hairy ape, hacking things out with hairy paws, and walking about on all fours. Theologians were aghast and some of their reactions are described in the Epilogue 1.

Biologists do not find their ideas ready-made. One needs an imaginative leap and then to refine concepts gradually, perfecting them as and when new evidence becomes available. Darwin's leading idea, the possible evolution of species by natural selection, gave him the problem to solve. Without a new idea that has properly formulated the investigation becomes an aimless collection of data and the energy expended on it is often wasted. At the start of the investigation, the truth or falsity of the original idea can be immaterial. What matters is its vitality and whether it gives rise to useful research. Even ideas that later prove to be unworkable have led to fruitful fields of enquiry. Alchemy led to the search for the transmutation of base metals into gold and was a great stimulus to the rise of chemistry; the search for a perpetual motion machine gave rise to a fuller comprehension of the interrelationships of energy in all its forms.

The evidence that Darwin published in his book supporting the idea of natural selection as a basis for the origin of species was overwhelming. He concluded that the forces acting on animal or plant populations (such as competition for food or living space, resistance to disease, or climate change) favored the survival and reproduction of those groups best adapted to the environment, a process he called natural selection. But the meaning of the phrase "natural selection" was distorted in curiously ingenious ways by opponents who were manifestly ignorant of what they were talking about. Darwin was attacked in both the press and the pulpit by such people. It was a striking example of the obstructions raised against the acceptance of a new idea. Plain facts can be apprehended in a moment, but a new idea is quite another matter. Acceptance requires an alteration in the attitudes of the whole mindset that was repugnant and even painful for many people upholding the Christian faith. However, Galton assimilated the contents of the *Origin of Species* as fast as he could read them; it gave him an exhilarating sense of freedom from theological bondage and dogma; it made him reassess man's place in nature and society from the viewpoint of the disadvantaged and impoverished.

Another strong proponent of Darwin's *Origin of Species* was Thomas H. Huxley[11] (1825–1895). Darwin first met Huxley at a Geological Society meeting in 1853 and was very much impressed by him. So was Galton when he heard him give a lecture at the Government School of Mines in Piccadilly. The young 28-year-old Huxley was up on the lecture platform performing intellectual gymnastics with strength and vigor. His face, bold and honest, was the face of a man who knows what he likes and knows what he hates. He was well aware of his strengths and it gave Galton real pleasure to see that there were such men in London at the time. Huxley would toss back his hair with a vigorous hand, those thick black locks similar to a lion's mane that hung down to his shoulders, and his eyes would burn into you as if you were the only person present in the room. Knowing of Huxley's interest in marine animals, Darwin with an impulsive gesture of friendship offered Huxley his collection of sea squirts from the time of the Beagle expedition to study and write up. This gift greatly pleased and flattered Huxley.

After a considerable struggle from an impoverished family background Huxley had by 1853 obtained a paid lectureship at the Government School of Mines. As a naturalist for the government's Geological Survey he became an expert in animal fossils and he was soon appointed to the prestigious position of Fullerian Professor at the Royal Institution.

Huxley reviewed a newly published monograph by Darwin on barnacles. He praised Darwin as a brilliant observer of nature on the small as well as the large scale, and he wrote that *it is one of the most beautiful and complete anatomical monographs, which has appeared in our time.* Darwin was delighted with this and his correspondence subsequently changed from opening letters with "My Dear Sir" to "Dear Huxley."

In April 1856, just 3 years before publication of the *Origin of Species* Darwin invited Huxley to visit his house at Downe where he had organized a small meeting for several other naturalists to sound out their views on the evolution of species. Darwin wanted to find out how the young Turks of the day would react to his ideas. At the time, Huxley was not particularly impressed, he was being more interested in the structural features of the animal world, the anatomy and the geometry of biological structures rather than how they change with time.

He was not yet really committed to the idea of the "evolution of species." Nevertheless, Darwin was gratified to note that some months later the notion of progressive development and modification of species was beginning to creep into Huxley's published articles and lectures.

Huxley had no idea what this meeting was about and was even rather reluctant to attend. Darwin instructed him to take the train from London to Orpington, and then Darwin would send his personal pony trap for the drive of four miles to the house at Downe. Huxley took the afternoon train passing the swarming streets of South London with its sooty foundries, and the great bellied chimneys tipped with smoky heat. At last he reached the peaceful fields and meadows of Kent. Darwin's pony trap was at Orpington station to meet him and take him to Down House, a rather ungainly and rambling mansion set in the meadows of Kent. The house served Darwin as his own Department of Biology to save him commuting to London and meeting obstreperous colleagues.

Huxley suited Darwin very well. The young man was an aggressive and fiery controversialist. He was spoiling for a fight against the established order of almost everything in England at the time. Anglican churchmen and the old order of naturalists were to be his obvious opponents. One eminent naturalist, Sir Richard Owen (1804–1892) was a particular target. Sir Richard was then superintendent of the Natural History Department of the British Museum, later to become the Natural History Museum, and was a distinguished comparative anatomist, perhaps the best in the country. Owen held some obscure views based on the Platonic ideal of archetypal forms being the design from which all animals were constructed. These were sort of divine blueprints on which the different animal classes (birds, reptiles, fish, and mammals) were built. He came in for a good deal of mauling from Huxley's pen: that his archetypal forms were ridiculous, his notion that the vertebrate skull was derived from modified vertebrae was absurd (but true), and that his system for animal classification was ludicrous. Even Darwin was astonished at Huxley's boldness and venom.

After the meeting Darwin came to believe that Huxley was a wonderful man, a veritable *enfant terrible* of the new biology. He sent Huxley parts of his big book on the *Origin of Species* asking for advice and verification of some of the finer points such as the time of first appearance of specialized bodily organs during development. Darwin said of Huxley: *When I felt myself chasing wild geese you always rise before me.* Darwin hoped that Huxley would provide major support for his revolutionary views on the *Origin of Species.* When his book came out in 1859 Darwin told his friend Joseph Hooker that he longed to hear what Thomas Huxley would think about it. In fact, Huxley's book on oceanic hydrozoa came out within weeks of Darwin's book but was generally forgotten in the excitement of the publication of the *Origin of Species.* The clergy queued up in droves to attack Darwin's book and Huxley generously rose to its defense. Darwin needed a defender for his ideas as much as Huxley needed a cause to fight. Huxley wrote that *the clergy decry it ... bigots denounce it with ignorant invective, and even savants ... quote antiquated writers to show its author is no better than an ape himself.* Huxley wrote two brilliant reviews of Darwin's book in the *Times* and the *Westminster Review.* Darwin was delighted; and from then on Huxley became Darwin's *agent provocateur* to promulgate the heresy of the "evolution of species," including man. Huxley's support culminated in a small book called *Mans Place in Nature* published in 1863. Its frontispiece was sensational. It depicted a *danse macabre* of skeletons, leading off with the gibbon and going on through to the chimpanzee, gorilla, and finally to man. Then followed provocative chapters on: Man—Like Apes; Relations of Man to the Lower Animals; and Fossil Remains of Man. Similar to Copernicus who had moved the earth from the center of the universe, so Huxley had moved man in his book from the center of animal creation to the periphery of an evolutionary tree of life, with just one branch reserved for the great apes where man was at a tip. By now, Huxley had fully adopted Darwin's views and was one of the most loyal disciples of Darwin's theories. His publication was nicknamed the Monkey Book by opponents and was usually to be found on sale among the more obscene and pornographic books at the larger railway stations.

Joseph Hooker and Thomas Wollaston were also among the party at Down House. Wollaston was an old Cambridge friend currently spending his time classifying beetles in the natural history section of the British Museum. The botanist Joseph D. Hooker (1817–1911) was Darwin's great colleague and supporter. He was quite a different character from Huxley, being the son of a famous father William Hooker, the Regius Professor of Botany at Glasgow and the first official director of the Royal Botanical Gardens at Kew. To keep a pure line of botany within the family, Joseph married Frances Henslow, daughter of another famous botanist John S. Henslow (1796–1861), Regius Professor of Botany at Cambridge (the same Henslow who recommended Darwin for the Beagle voyage). Joseph Hooker had six children but it was a great disappointment to him that none of the little Hookers ever became botanists. They did well in the civil service, in colonial administration, in engineering, and soldiering but none did pure science. Joseph became assistant director of Kew in 1855 and then succeeded his father as the director. Father and son were virtually synonymous with English botany for most of the nineteenth century.

Joseph Hooker was first introduced to Darwin in 1839 in Trafalgar Square after they had both been visiting the National Gallery. By the middle of the 1840s, they were on very friendly terms, and, just as with Thomas Huxley, Darwin had offered him a part of the Beagle collection, the Galapagos plants, to classify and write up. Hooker published at least two important articles on this material in the *Transactions of the Linnean Society* in 1851. It always worried Galton that Darwin never offered any of his Beagle collections to him; he consoled himself with the thought that he had never claimed to be a naturalist. In fact, he always found botany rather boring and used to get into a great muddle about the roles of stamens, pistils, carpals, and sepals.

Galton first saw Hooker in a painting done in 1849. It was called *The Great Botanist* in Sikkim. And there Hooker was seated against a background of the snow capped Himalayan peaks with local plant collectors and women kneeling at his feet handing the great collector sprigs of variously colored rhododendron blooms, while two servants were standing guard behind him. Hooker was dressed in a kind of multicolored striped dressing gown over a smart white suit and wearing a funny striped skullcap, looking rather regal and pompous.

In fact, on meeting him later Galton found him less pompous than he had imagined from the painting; he was quite a severe and dignified person—but always a trifle nervous. Unlike Thomas Huxley he would never make a good public speaker. He did not seem to have any great interest in appealing to a wider public, although his views on education, especially botanical education, were very sound. His disposition was too nervous and highly strung for him to stand out as a really effective public figure. He was 8 years older than Huxley but otherwise they had a lot in common. Their friendship lasted more than 40 years and strengthened considerably with time. They both became President of the Royal Society and their names stand next to each other in the list of people receiving the Darwin Medal.

It was their friendship and support for Darwin that drew them together. Hooker became Darwin's closest friend. They found each other's company very congenial and stimulating. They frequently corresponded and Darwin used Hooker as an authoritative source of botanical information and criticism of his pending works and theories of evolution. Hooker used to visit Downe frequently and often stayed for up to 10 days at a time at the house. He dedicated his botanical book *Himalayan Journals* to *Charles Darwin from his affectionate friend, Jan. 12th 1854*.

Galton's book; Hereditary Genius[12] 1869

Darwin's book set Galton thinking about the central topic of heredity: What is the mechanism whereby parents transmit to their offspring all the inherited characteristics? Galton was struck by the fact that some families appear to have many more gifted members than one would expect by chance. He collected the details of many of such families for his book *Hereditary Genius*. A good modern example would be the Huxley family. There was T. H. Huxley (1825–1895), an up-and-coming biologist in Galton's day and an aggressive controversialist defending Darwin; then his three grandsons Sir Julian Huxley (1887–1975), a brilliant zoologist who contributed to the modern synthesis of Darwin's theory and was also appointed the first director general of UNESCO (1946–1948); then there was Julian's brother Aldous Huxley (1894–1964), an internationally famous novelist writing *Brave New World* (1932) and *Eyeless in Gaza* (1936) among others; and finally Andrew Huxley (1917–2012) who shared the Nobel Prize for physiology and medicine in 1963 for his work on the conduction of the nerve impulse and muscle contraction. Galton came to believe that there was an inherited factor for mental abilities running through such families. He later admitted that he had wished he had titled his book *Hereditary Abilities* and not used the word Genius.

What was missing in Darwin's book was an explanation of how rare slightly beneficial variations in bodily structure or function in members of one generation were transmitted to the next. These beneficial variants were assumed to accumulate over succeeding generations and improve the fitness and reproductive ability, so that some members of Species

A could evolve into Species B under the influence of natural selection. Without these beneficial variants natural selection could not act as a driving force for evolution of species to occur.

Galton started his studies by attempting to explain the inheritance from parents to children for such characteristics as height, skin, and eye color, or liability to develop such diseases as asthma or pulmonary tuberculosis. He considered himself to be *a surveyor of a new country and to endeavor to fix in the first instance as truly as I could the position of several cardinal points.* In his view, the cardinal points were to define the disease carefully, to count the appearance of the disease in groups of identical and nonidentical twins as a measure of the extent of inheritance, and then to devise statistical tests for the prediction of inherited features. He argued that since identical twins have inherited an identical set of inherited factors, any disease, if inherited, should be found more frequently in these types of twin pairs than in nonidentical twin pairs that only share about 50% of their inherited factors in common.

Before studying twins, he collected 160 families with as many as 75 members in a single family to construct as complete a family record as was possible. This yielded approximately 12,000 disease states and 2,000 causes of deaths—he expended £500 of his own money in collecting this material. He demonstrated that his collected data were free of sampling bias by showing that deaths due to tuberculosis or suicide did not appear either more or less frequently in his records than in the ordinary Life Assurance Society mortality tables. He admitted that his observations were "slender," but considered the general approach of knowing the frequencies of a particular disease in the offspring of affected parents may give a value for the strength of inheritance for that particular disease. However, when he came to the final analysis of his family records he had to admit that he obtained practically nothing of value from the study. The records were too fragmentary and attempting to work with incomplete life histories could lead to serious errors.

He was more fortunate with his study of twins. There was a previous large scientific literature relating to the anatomical and physiological aspects of twins, but before Galton, there were no attempts to measure the psychological characteristics of twins. He studied 80 pairs of identical twins and 20 pairs of nonidentical twins. From his 80 pairs of identical twins he obtained 35 case histories suitable for analysis. The twins of seven pairs both had psychological disorders such as depression, paranoid delusions, hallucinations, or mania. Conversely with the 20 pairs of nonidentical twins he was struck by the dissimilarity of such case histories between the twin pairs despite at least 13 of these pairs having very similar family and educational backgrounds. Although much of the information he collected was anecdotal, he found with regard to the occurrence of disease that *there was no escape from the conclusion that nature (i.e., heredity) prevails enormously over nurture* (i.e., environmental conditions). The proof of such complex inheritance must, in his opinion, finally rest with statistical rather than with anecdotal evidence. To what extent does factor A in a parent contribute to factor B in the offspring? His answer was that *we must endeavor to find a quantitative measure for this degree of partial causation.* To this end, he developed the statistical test now known as correlation analysis.

He then went on to choose one of the thorniest of problems: the inheritance of mental abilities (intelligence tests had not yet been invented) in humans. Galton had been immensely impressed by the many obvious examples of inherited features among very able scholars at the older universities (his wife's family being one of them) and he determined to start on the subject of inherited abilities found in such families. He argued that reputation in the world was an approximate measure of mental ability. You had to be intelligent to make a name for yourself. His approach was therefore to make a list of all the famous men whose biographies he could find, mainly those published in the obituaries of *The Times* newspaper for the year 1868. He chose judges, statesmen, literary men, poets, musicians, painters, scientists, wrestlers, oarsmen, and so on. The lists were drawn up without any bias on his part because he always relied on the judgment of others. He then devised a system to grade the people in the list by their abilities and then constructed their family trees to establish how many close relatives were equally distinguished in some way or other. He worked out the data statistically, analyzing the percentage of eminent relations that were found in each family selected on the basis of one famous man. The tabulated results were very striking. They seemed amply sufficient to answer the main question that mental ability is certainly an inherited component of one's personality. Various objections as to the validity of his conclusions were made such as the influence of social upbringing and cultural environment in the presence of intellectually able compared to less able parents. However, Galton thought he easily disposed of such objections in research articles he published such as *Heredity in Twins*; *Typical Laws of Heredity*; and *Chance and Its Bearing on Heredity*. His chief regret was the choice of book title *Hereditary Genius*. There was not the slightest intention on his part to use the word "genius" in any technical sense, but merely to signify an ability that was exceptionally high and at the same time inborn.

His book came out in 1869, 10 years after the *Origin of Species*. It made its mark, although it was not nearly as influential as Darwin's book. The verdict that he most eagerly awaited was of course from his cousin whose letter, when it came, made him very happy.

> *Down, Beckenham, Kent 3rd. Dec.*
> *My Dear Galton—I have only read about 50 pages*
> *of your book (to Judges), but I must exhale myself else*

> *something will go wrong with my inside. I do not think*
> *I ever in all my life read anything more interesting and*
> *original—and how well and clearly you put every point.*
> *George, who has just finished the book and who expressed*
> *himself in just the same terms, tells me that the earlier*
> *chapters are nothing in interest to these later ones. It will*
> *take me some time to get to these latter chapters as it is*
> *read aloud to me by my wife, who is also much interested.*
> *You have made a convert of an opponent in one sense, for*
> *I have always maintained that excepting fools, men did*
> *not differ much in intellect, only in zeal and hard work;*
> *and I still think this is an eminently important difference.*
> *I congratulate you on producing what I am convinced*
> *will prove a memorable work…*
> > *Yours most sincerely, Ch. Darwin.*

Looking back over this letter, it may have been a trifle patronizing that Darwin only bothered to read the first fifty pages of his book before writing the letter, especially as he said it was so interesting and original. However, Darwin was now a world famous scientist, whereas Galton was still a relatively young man and still to find his way. Any praise from Darwin greatly flattered Galton.

One other thought struck Galton most forcefully was about their books. There was a vast field of ignorance in both their ideas about how inheritance actually worked. How were inherited features transferred from parents to their offspring? What were the biological mechanisms underlying this transmission? It was about this issue that their subsequent quarrel originated.

chapter three

Darwin's grand theory[13]

To ask or search I blame thee not, for [Nature]
Is as the Book of God before thee set
Wherein to read his wondrous Works, and learne.

Paradise Lost. J. Milton (1608–1674)

In 1865, Thomas Huxley (1825–1895) received a letter from Down House.

> *My dear Huxley,*
> *I write now to ask a favor of you, a very great favor from one so hard worked as you. It is to read a thirty-page manuscript, excellently copied out and give me not a long criticism, but your opinion whether I may venture to publish it in my forthcoming book on Variation in Animals & Plants Under Domestication (1868). You may keep the manuscript for a month or two.*

When they had last talked Huxley thought *Origin of Species* had serious flaws in the argument. Evolution of species according to Darwin depends on the accumulation of small variations in bodily structure or function that improves fitness for survival and reproduction. This gradual accumulation of new characteristics separates the animal or plant from the original parent species. Each of these small variations has to be transmitted to the next generation for them to accumulate there. How does this transmission work? This was a key issue to resolve if evolution by natural selection of minor variants of animal structure or function were to provide survival value in the long term.

Darwin's new idea about inheritance was that all the cells of the body throw off minute granules, which are dispersed throughout the organism. He called them "gemmules." They are collected from all parts of the body to accumulate in the sex organs, and from there they are transmitted to the next generation for the development of the new offspring; they are likewise capable of transmission in a dormant state to future generations and may develop at a much later time.

It became a passion for Darwin to try to connect all such facts by a single unifying system of thought or hypothesis, which is the general aim of all scientists. This was justified for him by the fact that it could correlate a large number of disparate single observations, and it is just here that the "truth" of a hypothesis should lie. The manuscript that Darwin wished to send to Huxley gave such an hypothesis. It was very crude as yet, but had been a considerable relief to Darwin's mind because he could hang many separate groups of facts on it. He well knew that this was a mere hypothesis, and this was nothing more, was of little value until tested by an experiment, but it was very useful to Darwin by serving as a centerpiece and summary of many chapters of his big new book on *The Variation of Animals and Plants under Domestication* (1868). Now, he earnestly wished for Huxley's verdict given briefly as "burn it"; or the most favorable verdict that he could hope for: *it does rudely connect together certain facts and I do not think it will immediately pass out of mind.* If Huxley could say this much and did not think it absolutely ridiculous Darwin would publish it.

He ended the letter:

> *You must refuse if you are too much over-worked; but I value your opinion most highly—you are so terribly sharp-sighted and so confoundedly honest. If you can read the manuscript be so kind as to suspend your judgment until you have read the whole and then turn the subject a little in your mind. I have thought of it much, more than appears in the manuscript and am becoming convinced that some sort of view will have to be adopted. The style of the script has to be improved. But I must say for myself that I am a hero to expose my hypothesis to the fiery ordeal of your criticisms. But you will be doing me a very great service.*
>
> *With cordial thanks, believe me yours sincerely, Ch Darwin.*

Darwin received a reply by return of post:

> *June 1 1865 Jermyn Street*
> *My dear Darwin—Your MS. reached me safely last evening.*
> *I could not refrain from glancing over it on the spot, and I perceive I shall have to put on my sharpest spectacles and best thinking cap. I shall not write till I have thought well on the whole subject.*
> *Ever yours, T. H. Huxley.*

The next letter that Darwin received from Huxley was lost, but from what Darwin reported it must have read something like this:

> *Do not publish. It is much too complicated and speculative. It makes such tough reading that perhaps I should come to Downe to give you my comments in person.*

Darwin initially swallowed the bitter pill as best he could.

> *My dear Huxley,*
> *I thank you most sincerely for having so carefully considered my manuscript. I do not doubt your judgment is perfectly just and I will try to persuade myself not to publish. The whole hypothesis is certainly much too speculative; yet I think some such view will have to be adopted when I call to mind such facts as the inherited effects of use and disuse of body parts etc. A visit from you would give me real pleasure. You know that the Orpington Station on the S.E. Railway is now open and only about four miles distant from us. I shall let you have no peace till you give me a full account of your objections to my hypothesis.*
> *Ever yours truly. Ch Darwin*

Darwin would usually greet guests on the front porch of his house probably dressed to go for his constitutional walk in his great coat. Ever since his marriage Darwin had suffered from frequent bouts of ill-health. He led a most regular life since then, breakfasting alone at about 7.45 a.m. and working afterward for about an hour and a half till 10.00 a.m. He would then come into the drawing room for his letters and if there were any family letters they would be read aloud to him as he lay on the sofa, frequently wrapped in a shawl. From 10.30 a.m. to about noon, he would work again, after that he would consider the work done for the day and would go walking with his dog, wet or fine. His bouts of ill-health at this time were mainly from indigestion and insomnia and periodically he would go away to take the waters at some spa town in the hope of a cure.

Today he could take Huxley on a well-beaten path—the Sandwalk. This was a quarter mile circuit set up by Darwin on the property behind the house that he had rented from his wealthy neighbor, Sir John Lubbock, the banker. The path went right to the furthest end of the kitchen garden and then through a wooden door in the high hedge. Here a fenced path ran between two large lonely meadows until you came to the wood. The path ran straight down the outside of the wood and Darwin usually walked this path three times daily for his constitutional; he planted it with some shrubs and trees and found it a very pleasant place to walk, to think about his ideas, or to discuss matters with his friends.

The conversation would most likely run on these lines. Huxley did not want to stop Darwin from publishing his views. Huxley really would not like to take that responsibility. Somebody rummaging among Darwin's papers half a century later would find Darwin's hypothesis and say *See this wonderful anticipation of our modern theories—and that stupid ass Huxley prevented his publishing them.*

Well what did Huxley have against the theory?

As Huxley understood it, Darwin's hypothesis proposed that every organ of the body gives off a stream of hereditary particles, which Darwin called "gemmules." Darwin postulated that these particles or gemmules are shed from all the organs of the body: the brain, the heart, liver, kidneys, the muscles, and so on in great numbers and have the power of multiplying and propagating. Each gemmule does not contain a microscopic blueprint of the complete organ or individual as Herbert Spencer (or even Aristotle before him) believed, but only carried certain characteristics of the organ of the body from which they came. They circulate to the reproductive organs and accumulate there. They seemed to possess a marvelous elective power by which they can all assemble together in the proper place and in due proportions. A certain number are then incorporated into the sperm or egg in proportion to the numbers that arrive from the various body parts. It was these gemmules, or hereditary units, that determine the inherited features of the offspring. So, each offspring would receive a different collection of gemmules from their parents, some from the father mixing with others from the mother; and it was these different mixtures that accounted for the differences between brothers and sisters. It formed the basis of a blending type of inheritance.

Darwin called it his Pangenesis or Gemmule theory to imply that all the body organs (Greek *pan*—all) can generate (Greek—*genesis* create) the hereditary material or the gemmules. It is as though every cell in the body has a "vote" in the hereditary constitution of the offspring. It explains so many things. For example, when a white man marries a black woman their children are invariably colored brown; that is because equal numbers of skin gemmules from each of the parents are transported to their sex cells (egg or sperm) and they mix or blend together in their child to produce the brown intermediate skin color.

In that case, how could any new rare variant ever become established in a population? Suppose the rare new variant is likened to a small drop of black paint and it is added to a bucket of white paint (being the property of the gemmules coming from the normal organ), the variant color black would vanish immediately when mixed (or bred into) into the white paint and the variant would be lost forever. In a similar manner, the rare variant gemmule would soon be diluted out by all the normal gemmules of the particular organ of the usual breeding population. So, blending inheritance cannot be a major mechanism for producing variation on which natural selection depends. Darwin could not provide an answer to this objection.

Or again if you think of eye color, the marriage of a brown-eyed mother to a blue-eyed father can produce a child with brown eyes; it does not have to be an intermediate color. Therefore, blending does not appear to be the case here; inheritance appears to be more discontinuous. Here, Darwin had to suppose that all the gemmules from the father were impaired in some way and could not come through strongly when mixed with the mother's gemmules.

However, Darwin's hypothesis could explain how disuse of a body part becomes inherited. Consider those fishes that live deep in the ponds and rivers found in the caves of Carniola (Central Slovenia) and Kentucky. Their ancestors appeared to have had eyes; but in the cavefish they have become vestigial and the fish are quite blind. The same occurs in cave insects and small cave-dwelling mammals. Their eyes liberate fewer and fewer gemmules over the years as they atrophy; and so the eye is less and less represented in the sex organs and eventually no eye gemmules are transmitted at all to succeeding generations. So, they become blind—an example of disuse inheritance. Huxley pointed out that this was getting perilously close to Lamarck's scheme for the inheritance of acquired characters. This annoyed Darwin excessively, he hated to be identified with these ideas of Lamarck (1744–1829). He wanted nothing to do with the idea that the environment creates bodily structures; that animals strive to better themselves by adapting to their external surroundings and then pass on these adaptations to their offspring. Lamarck's book on this, *Philosophie Zoologique* (1809), was a wretched work from which Darwin claimed he had gained nothing. It is too ridiculous to suppose that just because the giraffe liked to browse on leaves near the top of trees that are tall, it managed to grow a longer neck to reach them; and this long neck could then be passed on to their offspring. Darwin would propose that a chance variation in the ancestor of the giraffe gave rise to a longer neck; and that these longer-necked animals were more successful in the struggle for survival over shorter-necked animals for obtaining their food supply at the top of trees and therefore would have a survival (and a reproductive) advantage. Quite a different process from Lamarckism; but both Darwin's ideas and Lamarckism face the problem of dilution out of the slight beneficial advantage in future generations by blending inheritance.

Huxley went on to point out that disuse of a body part does not seem to be inherited in all cases. Take for instance, circumcision in Jews; this has been going on for thousands of years. Their sons must be getting no foreskin gemmules at all from the father, yet their sons are all born with foreskins.

Darwin could argue that the timescale of circumcision had not been going on long enough for it to show as an inherited character but may do so with time. Since gemmules were transmitted over many generations, the removal of a part posed no problems because the gemmules derived from the lost part still persist in the sex organs and are multiplied and transmitted from generation to generation. However, with the passage of time without replenishment, they would eventually be expected to die out.

Darwin's theory could also account for the regeneration of body parts after injury. If the leg of a newt is cut off, it regenerates a new leg in an exact proportion to the size of the adult animal. There are only mature cells present at the site of amputation, and these could not possibly grow into a new leg.

But the gemmules at the injury site, attracted by the severed surface of bone, muscle, and nerve pass into or, as it were "infect" the mature cells of each tissue and allow them to multiply till the whole limb is exactly reproduced.

However, this does not always happen. If you amputate a finger from your hand, it does not regenerate a new finger despite the presence of gemmules here. Why is that? Darwin again could have no answer for this.

Huxley came up with other objections. The gemmule theory does not explain sudden changes in body structures; such as the appearance of a cleft lip or cleft palate. These can occur in children with no history of the condition in either parent; the deformity can also appear in the subsequent children of the affected adult. How does the gemmule theory account for such deformities?

Darwin did not know but provided a weak explanation. Some gemmules lay dormant and only became known under certain circumstances. So gemmules for the new cleft palate were already present, although latent in one of the parents.

In addition, take this phenomenon of skipping a generation; a white child can be produced from the marriage of a mixed-race father and a white mother. If the child then comes to marry a white partner, some of their children can show colored features again. They show the characteristics of the grandparent but not the parent. So, where do the "black" skin gemmules come from in this case? That was easier to explain on the basis of latency—the white child always harbors a few black gemmules in the sex organs that lie dormant, just like seeds do in the ground. The gemmules can then become activated at a later time by some environmental stimulus and become incorporated into eggs or sperm for transmission to the child.

Huxley appeared doubtfully convinced. The whole idea seemed to hark back to Hippocrates (c.460–c.377 BC). This ancient Greek physician postulated, unlike Aristotle, that the reproductive seeds come from all parts of the body (heart, lungs, kidneys etc.) and the seeds travel from all the organs to the brain, where they travel down the spinal cord to the sex organs for onward transmission to fuse with seeds of the opposite sex at the time of reproduction to form the embryo. Incidentally, masturbation in boys was later thought to lead to feeble-mindedness or even madness, because one's brain would leak away with the excessive spilling of the seeds in the semen.

Hippocrates in his book *The Sacred Disease* (referring to the "falling" sickness or epilepsy) that like most other diseases epilepsy is hereditary and not visited on humans by the Gods as a punishment for sin (translated by Chadwick and Mann 1950). The brain is the seat of the disease and if the seed coming from this organ is defective the children may suffer from epilepsy. He also gave a not implausible reason for his time of why the disease develops due to an imbalance of the four body humors: blood, phlegm, yellow bile, or black bile, with too much phlegm occurring in epileptics to block the arteries supplying the brain. It was a great step forward for its time to think epilepsy was not of divine origin but had a rational explanation for its occurrence. Body humors do occur; they can be measured, and are more rational entities than such metaphysical suppositions as phlogiston postulated by the early chemists or vitalism to account for the chemistry of life by biologists.

Later Darwin was to admit that he wished he had known earlier about the views of Hippocrates as they seemed almost identical to his own—merely a change of terms and application to classes of facts necessarily unknown to the ancient Greeks. The whole case was a good illustration of how difficult it is to come up with anything really original.

Despite the deficiencies that Huxley pointed out, Darwin still thought his hypothesis as an important step forward in biology. No doubt the whole gemmule hypothesis was complicated—but then so were the facts. It may have gaps, although after mature reflection Darwin believed that biologists would someday be compelled to admit some such mechanism. After all, there was a letter to Charles Lyell (the leading geologist of Victorian England) from Alfred Wallace (1823–1913, the codiscoverer of natural selection) that:

> *The hypothesis is sublime in its simplicity and the wonderful manner in which it explains the most mysterious of the phenomena of life. To me it is satisfying in the extreme. I feel I can never give it up unless it be positively disproved, which is impossible, or replaced by one which better explains the facts. Darwin has here decidedly gone*

ahead of Spencer in generalization. I consider it the most
wonderful thing he has given us, but it will not be gener-
ally appreciated.

Wallace also wrote directly to Darwin:

I can hardly tell you how much I admire your hypoth-
esis. It is a positive comfort to me to have any feasible
explanation of a difficulty that has always been haunting
me—and I shall never be able to give it up till a better one
supplies its place … and I think that hardly possible.

Darwin agreed with this wholeheartedly and wrote:

To my mind the idea has been an immense relief, as I could
not endure to keep so many large classes of facts all float-
ing loose in my mind without some thread of connection
to tie them together in a tangible method.

So, Darwin would accept Wallace's opinion and reject Huxley's. Go for publication? Or better still get another opinion. Perhaps Joseph Hooker would tell Darwin what he thought of it.

For the rest of the stay at Down House, Huxley was not comfortable especially during meals. He recognized that Mrs. Darwin was a remarkable woman, although she always appeared rather stern and forbidding. It was so different from his house in Marlborough Place where his young children were cared for by Mrs. Huxley in an exceedingly kind and pleasant manner. The whole domestic tone of his house made visitors feel quite at ease. Alfred Wallace used to visit the Huxleys regularly on Sunday afternoons and he said that it was one of the rare houses where he was made to feel perfectly at home.

Mrs. Darwin on the other hand was mainly preoccupied with all the illnesses of her children and in the meantime acting as a devoted nurse for her husband. The children in many ways played up to this atmosphere of hypochondriasis and seemed to imitate their father in all his symptoms. In fact, Mrs. Darwin was extremely tender hearted and rather too sympathetic for her children's best interests when they were unwell. A little neglect and astringency might have done some of them a world of good. The children seemed to treat their father as a sort of cross between God the Father and Father Christmas. They could play informally with him up to a point, but he often reverted to becoming like an all-powerful God. Huxley thought that some of their ailments might have been of nervous origin, that there was a certain mournful pleasure in being ill and then to be petted and nursed by their mother. The attitude of the whole Darwin family to sickness was unwholesome, and this persisted with the children into an adult life. Their daughter Henrietta remained a hypochondriac for the rest of her life. She was always going away to rest in case she might get tired later in the day, or even in the next day. She would ask the cook to count the prune stones left on her plate as it was very important to know whether she had eaten three or four prunes for luncheon. When there were colds, she invented a kind of gas mask. It was an ordinary wire kitchen strainer stuffed with antiseptic cotton wool and tied across her nose like a dog's snout, with rubber bands over her ears to hold it on. In this, she would discuss politics in a hollow voice out of her eucalyptus-scented muzzle. She characteristically wrote to a proposed visitor: *"Don't come by the ten o'clock train, but by the 3.30 so as to give me time to put you off if I am not well."*[14]

After Huxley had left the next day to catch the train back to London, Darwin went through all the arguments against his ideas. He had to recognize the force of some of them and thought he should follow Huxley's advice and seek another opinion from his friend Joseph Hooker. If Hooker came out strongly against it, perhaps he should drop the idea or at least try to modify it. A few days later he set about writing to explain the position.

> *My dear Hooker,*
>
> *I do not know if you have met Thomas Huxley recently and whether he told you about my new hypothesis. I think this one is quite as important for the future of Biology as my ideas on Natural Selection have already been. However Huxley did not like it and suggested that I come to you for your opinion. I should very much like to hear these since I value your views above any other man in England due to the course of your own studies and for the clearness of your mind. I do not know whether you have ever had the feeling of having thought so much about a subject that you have lost all power of judging it. This is the case I am in after 26 years of reflection, so I badly need your assistance. I enclose a fair copy of my hypothesis that I hope to publish in my next book.*
>
> *By the way there is a rather nice Review of you in the last Athenaeum; and a very unnice one of my book; I suspect from two or three little points that it is by Owen—he so despises and hates me.*
>
> *Ever yours very truly,*

Darwin did not have to wait long for the reply. As it happens, Hooker had met Huxley and they discussed Darwin's hypothesis extensively. He wrote that Huxley had made a very clever remark that was *so deuced clever that I cannot quite clearly recollect it, still less write it down ... but it was something to the effect that the cell might not contain gemmules but a potentiality in the shape of an homogeneous material, like a crystallizable compound that is present in various isomorphic forms, depending on some unknown influence. And this determines the pattern of inheritance from parents to offspring.*

Darwin probably chuckled to himself at his friend who after hearing such a clever remark could not recollect it, and then started out on an hypothesis even wilder than his own.

Hooker went on:

> *I do think your hypothesis is unnecessarily complicated. It is so very speculative and a thousand times more difficult to grasp than the Atomic Theory or Latent Heat of Evaporation. At least with the chemist's theories of atoms and molecules their imagery is useful. They convey definite ideas about atoms combining in certain strict and constant proportions to produce molecules. If Biology enabled us so to convey definite ideas about the combining properties of your gemmules they would have their use. But inasmuch as organisms are not given to unite in definite proportions I do not see what you gain by postulating them.*
>
> *I do not say that gemmules do not exist; they may even be like minutes of time, purely arbitrary quantities that do help us to understand the passage of time, so in the same way your gemmules may help us to understand the phenomena of inheritance. But your doctrine of gemmules emanating from cells in no way furthers my perceptions or advances my understanding. Also what really is new in your hypothesis? Parents do transmit something to their offspring, which you have chosen to call gemmules—that is all.*

Darwin read these comments carefully but was not really dissuaded. He believed that Hooker did not properly understand his notion of gemmules. He was reminded of the answers that an unusual boy gave who could do remarkably complex calculations in his head. When pestered by many mathematicians to tell them how he did it, he replied in exasperation that: *God put it into my head and I can't put it into yours.*

So, Darwin felt about gemmules and Hooker's head. Darwin thought that a helpful analogy was to consider that certain inherited characteristics of parents were "photographed," as it was, onto the child by means of material gemmules. These transmitted features then "developed" in the child similar to a photographic plate.

Darwin later told Galton that his hypothesis, as a picture or map of what really happens in the natural world, was likely to give as great a value and coherence to scientific thought and investigation as his earlier ideas on natural selection. He felt a deep conviction that the gemmule hypothesis would someday be generally accepted and would be looked on as the best idea to explain the nature of inheritance, the repair of injuries, and how the body develops from a single cell. In some future day, biologists would be compelled to admit some such doctrine. He published it in his big book on *The Variation of Animals and Plants under Domestication* (1868) that is 2 years after Mendel had published his ideas on heredity. Darwin wrote:

> *I assume that the cells of the body throw off minute granules which are dispersed throughout the whole system; that these, when supplied with proper nutriment, multiply by self-division, and are ultimately developed into units like those from which they were originally derived. These granules may be called gemmules. They are collected from all parts of the body to constitute the sexual elements, and their development in the next generation forms the new being.*

Darwin was always delighted to see a word in print in favor of his own ideas. He thanked Ray Lankester for saying a few kind words about it:

> *I was pleased to see you refer to my much despised child... who I think will some day, under a better nurse turn out a fine stripling.* And when John Tyndall spoke about gemmules in his presidential address to the British Association Darwin responded: *You are a rash man to say a word for my gemmule hypothesis for it has hardly a friend amongst Naturalists, yet after long pondering (how true your remarks are on pondering) I feel a deep conviction that it will some day be generally accepted.*

The proposition that the Earth is round was the only hypothesis in earlier times and many people thought it was flat. Now we have so much evidence in favor of its roundness that one would be considered mad to say it was flat. It will be the same with gemmules. And even in the year of his death in 1882, he was writing to one of his junior colleagues, George Romanes, wishing him every success in his experiments, to prove the validity of the gemmule hypothesis.

With Mendel's binomial hypothesis of inheritance, we have no idea how it originated. Which came first, the hypothesis and then his 8 years of experiments; or *vice versa?* Unlike Darwin it appears he had no one with whom to discuss his ideas about the project. Having someone to clarify one's thoughts can be a very great help. Mendel did exchange scientific views in at least 10 letters with Carl von Nageli but the first one was sent in 1866 and therefore after the completion of Mendel's main work, which arrived in the scientific community out of the blue. There is little evidence that von Nageli appreciated Mendel's results and did not refer to them in his own publications on heredity. The gist of von Nageli's correspondence with Mendel was "mistrustful caution." Mendel's ratios were only empirical and his interpretation of them had gone beyond what the data allowed. If Mendel were to be correct, then he, von Nageli, had to be wrong in his own theory about the idioplasm conveying hereditary influences. This gives rise to an uncomfortable feeling in a world renowned scientist and might have been the reason for von Nageli recommending Mendel to work on a different experimental model, the Hawkweed (*Hieracium*), a genus of the sunflower (Figure 3.1) and closely related to the dandelion. Mendel did so with negative results and as it turned out, this plant was entirely inappropriate for the type of experiment Mendel was doing because of its asexual means of reproduction. Its seeds were just clones of the parent plant. Mendel would therefore be unable to replicate his results with the edible pea of finding consistent ratios of inheritance in two carefully selected characteristics of the plant. This can be very discouraging for any scientist and makes one wonder if this is a special case for the edible pea, or perhaps his results were just plain wrong—as later claimed in 1936 by the British geneticist Sir Ronald Fisher (p. 143).

Figure 3.1 Hawkweed or *Hieracium auriculi*, a relative of the dandelion and sunflower, which Professor von Nageli recommended Mendel to study, which he did. Most hawkweeds reproduce asexually by means of seeds that are genetically identical to their mother plant. So, Mendel could never reproduce his observations that he found with the pea that requires sexual reproduction. (From Thome's catalog Flora von Deutschland, Osterreich 1885.)

chapter four

Trial by experiment[15]

Experiments escort us last
Their pungent company
Will not allow an Axiom
An opportunity.

Emily Dickinson (1830–1886)

The work on the gemmule hypothesis began with an unexpected letter from Galton to Darwin: *My Dear Darwin, I wonder if you can help me.*
I want to make some experiments that have occurred to me in breeding animals....
His idea was to find out whether the hereditary units or gemmules of a
black breed of rabbit could be passed by a blood transfusion into a white
breed, such as the Angora Albino, to see if the latter would then give rise
to mongrel offspring showing features of both breeds of rabbit. Was that
blood, as supposed by folklore the world over ("he is my blood relative"),
the true bearer of hereditary particles? The letter was to ask Darwin where
he might obtain Angora Albino rabbits. Galton (Figure 4.1) had rather
uncritically adopted Darwin's views about the inheritance of gemmules
and incorporated the ideas piecemeal into his book *Hereditary Genius*
(1869) in the chapter on General Considerations. He now seemed to want
to test the ideas experimentally. Darwin was obviously very much interested in such a project and they agreed to discuss it further when Darwin
was next up in town.

They occasionally met in the library of the Athenaeum when Darwin
had made one of his rare visits to London. Darwin hated coming up to the
city, the journey exhausted him, and he invariably felt unwell by the end
of the day. He liked to get his London business done in the early mornings
and then escape back to Downe as quickly as possible. As an eminent past
member of the club, there is now a very fine portrait of Darwin staring
down at members similar to an Old Testament prophet (Figure 4.2) from
the high wall of a ground floor room.

The Athenaeum, situated on the south side of Piccadilly, was and
still is a time-honored gentlemen's club. However, what was becoming
an increasing nuisance around the Athenaeum in the late nineteenth
century and around the adjacent streets leading up to Trafalgar Square
was the number of child street sellers. One would bump into a diminutive
child, often a girl, of about 7 years, dressed in rags and carrying a wicker

Figure 4.1 Francis Galton (1822–1911); from the Mary Evans picture library.

Figure 4.2 Charles Darwin "like an Old Testament prophet".

basket with small posies, selling them for about a halfpenny a bunch. She would look up with her pale sunken face and hollow cheeks, and when she began to speak, a coughing fit would often take over. Her total family income would be about 10 shillings per week, what William Booth (the founding father of the Salvation Army) called "vicious poverty." These children swarmed in the streets of London, pilfering what they could, and then went home to sleep in rotten tenements similar to crowded maggots. They drove the thoughts of heredity temporarily to the back of his mind; environmental issues of poverty, squalor, infections, and exploitation of children seemed to be the more urgent problem. Galton could not help thinking that it would be better if such children had never been born. What was the point of bringing children into the world condemned to poverty and a miserable, uncared for existence? Such ideas were to form the keystone of Galton's social work on "eugenics" that he developed later in the nineteenth century. What was this problem? It was to give a few thousand people enough to eat, decent houses, and a fair income. Britain could well afford this as one of the richest of industrial nations in the world.

When Galton arrived at the Athenaeum, he went through the grand neoclassical entrance with a copy of the Apollo Belvedere at the head of the sweeping staircase. He went straight up to the first floor to the quiet West Library. They met to discuss the problem that both their books left unanswered. How does inheritance actually work? Darwin's concept of natural selection needed inherited variants to provide the raw material for evolution to work on. And what is it that is transmitted from parents to children to account for the transmission of eye color, body height, or even mental abilities, and all the other inherited features of the gifted families Galton had written about? Darwin thought his gemmule hypothesis accounted for and explained most of the known facts. Darwin had been pondering over these ideas since the 1840s.

Galton certainly agreed that the exact nature of inheritance was one of the most pressing scientific problems of the day. To elucidate this accurately would be a very difficult task demanding an altogether unusual capacity for anyone attempting it. He had heard about Darwin's ideas on hereditary gemmules—so, where did they stand now?

Darwin told him that his theory was bound to be generally accepted someday. For him it had the hallmarks of a great idea, namely it was original he thought, it seemed economical of application, and had the ability to connect many other facts together. It had considerable support from Alfred Wallace and Sir Henry Holland, who had read it twice and thought that some such doctrine or one closely akin to it would have to be admitted in the near future. Darwin personally found it a great relief to have some definite and plausible view of how inheritance actually worked. There was also an undercurrent of anxiety and self-interest in all the Darwin's speculations on heredity. He always wondered whether his bouts of illness were hereditary and because he had married Emma, his first cousin, would he inadvertently pass on some of them to his children. Whenever any of his children fell ill, he always thought he could recognize some of his own symptoms in them. Perhaps the gemmule idea would allow him to believe that the risks of passing on his own weak constitution would be reduced by dilution of his disease carrying gemmules by different ones coming from his wife—this thought might have been a considerable comfort to him. He told Galton that both Huxley and Hooker did not like the idea—though Darwin was not sure that Huxley understood it properly. Huxley tended to rush into criticism before having maturely reflected on all the implications. Darwin was convinced that all the skeptics would be converted someday soon. Unfortunately, the skeptics were rather numerous at the time; and strong convictions can be more dangerous foes to the truth than outright lies. Galton now wanted to apply rigorous tests to Darwin's hypothesis by way of experiments to see if the idea would hold up. They would agree to set up a systematic enquiry to try to determine whether such gemmules exist. Galton would test Darwin's hypothesis to see if would hold water. But what exactly were they trying to find out?

Their goal would be to try to identify the nature of the material, with the greatest possible accuracy that passes from parents to offspring and determines the transmission of inherited features. Some features of offspring strongly resemble those of the parents; whereas others seem to be quite different. They should concentrate only on the shared features and study these in detail. The fur color of rodents can be strongly inherited and this would make an excellent experimental model. For example, if you mate albino mice together, their offspring are always albino; therefore, "albino" gemmules might be the easy ones to identify.

They made a plan, at least in outline, so that they knew which experiments they had to perform in order to hope for reasonable results. There is no point in doing experiments that do not have a chance of providing answers to the questions posed. They both considered experiments to be the keystone to doing science; the results can completely destroy one's most cherished hypothesis. In biology, crucial experiments are even more difficult to devise and to perform properly than in the other sciences such as physics or chemistry, because of the immense complexity of living systems. Difficulties of simplification and countless sources of error can arise due to inadequate definitions and improper formulation of the problem. However, they had found an interesting field to study and were determined to find a solution.

After considerable discussion, they agreed to do the following project together along these experimental lines. A statement of the experimental protocol would be as follows:

1. The hypothesis to be tested would be taken straight from Darwin's ideas as published in *The Variation of Animals and Plants under Domestication*. Different parts of the body are supposed to liberate hereditary units called gemmules that accumulate in the sex organs for transmission in the sperm or egg to the next generation. Such gemmules account for the inherited features in a family tree.

2. For the experimental design: they chose to investigate the color of the fur of rabbits because they believed it to be under hereditary control. Pure-breeding white furred rabbits would be transfused with blood coming from pure-breeding black furred rabbits (Lop-eared variety) to see if the coat color of the offspring of the white rabbits can be changed into black fur due to the infusion of the black rabbit gemmules; and then do the *vice versa* experiments.

3. The choices of materials and methods to be used in any experiment are usually decisive for its success, so they would try to keep them as simple as possible. They would use two inbred strains of rabbits to produce at least 50 progeny from white rabbits whose blood volume of both parents would be replaced by at least 20 percent with blood from the black strain of rabbit. The progeny of the white transfused rabbits would be studied and then mated again to produce a second generation of rabbits to see if their progeny would in turn be altered to resemble the color of the black furred rabbits that donated the blood.

4. The results would be expressed as the numbers of rabbits produced from the transfused white rabbits that show any evidence for a change in fur color to the black variety and compared them to the rabbits that showed no change in fur color despite the infusion of black furred gemmules.

It was the simplest of the experimental protocols that any student should be able to follow. The results, whatever they show would be submitted for publication in the usual way. They would also write them up in plain English, so that everyone could understand them. *You know how I strongly believe* Galton had often said *that we do not do science for ourselves, but we do it so that we can explain it to others in plain language.* They both agreed that clarity of thought leading to clarity of language is the least courtesy that they owe to their readers. That was why Galton often wrote an article complete with all the scientific jargon for the specialist journals; and then another in a popular magazine on the same topic written for the general public.

Darwin readily agreed to all this. But time was pressing and Darwin always wanted to make his way back to Downe as soon as possible. He was already tired and much later than usual in leaving London; he wanted to get home before nightfall.

chapter five

A bolt from the blue—Darwin's response

If you do not expect the unexpected, you will never find it.

Heraclitus (c.535 BC–475 BC)

Whenever Darwin returned home from one of his expeditions to London there was always a pile of posts awaiting him on the side table in the front hall; there were volumes of different learned journals done up neatly in brown paper parcels tied with string, some books that he had ordered; and the customary heap of letters from his admirers and detractors. He took the whole bundle of post into his study and methodically sorted out the letters.

When he opened a letter, it has been said he found a reprint of a previously published article from a Catholic monk working in the Augustinian monastery of a small city called Brünn, north of Vienna. He had never heard of the author (Figure 5.1) and the German title of the article in the 1902 English translation was:

> *Experiments in Plant-Hybridization by Gregor Mendel,*
> *read before the Society for the Study of Natural Science*
> *of Brünn in 1865* (using the 1902 English translation
> in ref. 3.1).[16]

Darwin was usually meticulous in assimilating new materials and making notes about it. He could read German only slowly, a few pages at a time, and found it a chore. He was already corresponding with several top European scientists who were working on the broad issue of heredity. They were Carl F. von Gaertner in Germany working on plant hybrids; August Weismann in Freiberg and the Swiss botanist Carl Wilhelm von Nageli, now working in Munich. Carl von Nageli also received a reprint from Mendel and subsequently exchanged letters with him over 7 years on the topic of heredity. Von Nageli's main subject was cell division and pollination but he achieved the doubtful distinction of the man who discouraged Gregor Mendel from pursuing further work on plant inheritance, perhaps because he had his own different theory of inheritance

Figure 5.1 Gregor Mendel aged more than 46 years. Note the pectoral cross showing that he was now the abbot of his monastery.

by the idioplasma (a part of the cytoplasm). Mendel had of course read, studied, and quoted from the *Origin of Species* in the German translation, *Uber die Entstehung der Arten*, as soon as the second edition appeared in 1863. In his personal copy, he penciled many notes in the margin with his small and careful handwriting with double underlines of some of the text and even interspersed with the occasional exclamation mark. He bought most of Darwin's other works and studied them carefully making frequent annotations. So, it would be natural for him to send Darwin, as an eminent English biologist, one of the forty reprints of his own work.

If Darwin thought German articles were important enough, he would get them translated by William Dallas, a competent naturalist who often prepared the index of Darwin's books. He did not do this for Mendel's article and it was not rendered into English until 1902 by the Royal Horticultural Society.

If he had read the article, and some scholars dispute this, he would have found that the aim of the work was to report experimental evidence for a mechanism of inheritance, which would enable a plant species A to evolve into a separate species B followed by five pages of his results explaining how this could happen. This alone should have excited Darwin's interest. Mendel writes in his introduction that he is *searching for a law governing the formation of hybrids* [...in the edible pea...], *to arrange the forms with certainty according to their separate generations, and to ascertain their statistical relations*. Statistics were necessary as he was using large populations of plants. The three clear aims that he adopted ([1] searching for laws; [2] connecting one generation to the next; and [3] using statistics

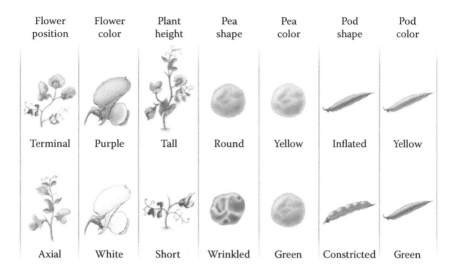

Flower position	Flower color	Plant height	Pea shape	Pea color	Pod shape	Pod color
Terminal	Purple	Tall	Round	Yellow	Inflated	Yellow
Axial	White	Short	Wrinkled	Green	Constricted	Green

Figure 5.2 The seven traits in peas on which Mendel eventually settled for his studies. (From Opitz, J.M. et al., *Ital. J. Pediatr.*, 42, 35, 2016. With permission.)

to analyze large numbers of plants) led to the whole success of his research project. He knew what to look for (Figure 5.2), and for his day these three features for an experiment of looking for regularities in hybrid forms, studying subsequent generations, and using statistics for analysis were absolutely new. He goes on to write that if he gets a correct solution to the problem its *importance cannot be overestimated in connection with the history of the evolution of organic forms.* It might seem odd that a Catholic priest was doing experiments to disprove the fundamental Catholic Doctrine of Creation whereby animal and plant species are formed by God and considered to be immutable.

The results of a series of experiments carried out for more than 8 years on hybrid formation in the edible pea (*Pisum sativum*, Figure 5.3) were then described. This appeared to require the patience of a saint involving the cultivation of a seven different lines of inbred pea plants (e.g., tall or dwarfed plants; with white or colored flowers; smooth or wrinkled peas; constricted or inflated pea-pods etc...), planting them in allotted beds, then relying on self-fertilization, which they do readily due to the enclosed shape of their flowers; or doing cross-fertilization by collecting the pollen from the stamens (the male part of the flower) from selected plants with a fine camel hair brush and dabbing it by hand onto the stigma (the female part of the flower) of other selected plants whose stamens had been previously removed with a pair of tweezers, so rendering the plant unable to self-fertilize. He prevented alien fertilization by

Figure 5.3 *Pisum sativum* from Thome's catalog Flora von Deutschland, Osterreich 1885. This was the plant on which Mendel spent most of his efforts. (From Opitz, J.M. et al., *Ital. J. Pediatr.*, 42, 35, 2016. With permission.)

using protective bags over the flower (Figure 5.4). He was very thorough by doing reciprocal crosses, that is, pollen of one plant type being transferred to the stigma of the other flower, and vice versa. Incidentally, he showed that each sex contributed equally to the offspring, which at the time was denied by some biologists.

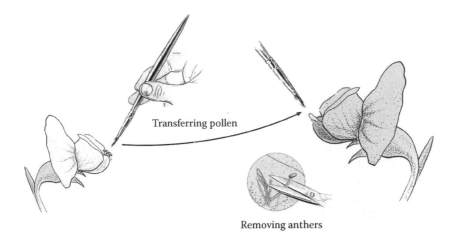

Transferring pollen

Removing anthers

Figure 5.4 Mendel did his cross-pollination in peas with a paint brush and forceps. A great advantage of this species is the ease with which it can be cross-fertilized. (From Opitz, J.M. et al., *Ital. J. Pediatr.*, 42, 35, 2016. With permission.)

The picture of Mendel dabbing the stigmata with yellow pollen dust with his paint brush brings to mind the picture of Seurat with his pointillism style of dabbing with his paint brush the yellow dots on his canvas to create the impression of sunlight in his scenes of Parisian parks 30 years later.

Mendel then claimed that he had discovered an explanation for the striking (almost mathematical) regularity in which the same hybrid characters, either the common or rare forms, appeared in some of the plant crosses in a ratio of 3:1. The author thought the importance of this whole number ratio could not be overestimated when it came to the history of the evolution of plants because he could calculate the exact numerical relationships for the appearance of many inherited characters of the edible pea across two or more generations. There are other ratios that are found in nature such as the Golden Ratio of 1:1.618 derived from the Fibonacci number sequence,[17] but Mendel's ratio bears no relation to this. It more relates to the fertilization sex ratio of 1:1 (random occurrence of son or daughter, nonblending inheritance, large samples needed to observe the 1:1 ratio etc.).

Mendel's article then went on section-by-section describing the best choice of an experimental plant to use and this turned out to be the edible pea. His patience was phenomenal; he spent about 2 years making sure that the features of the plant he was going to study bred "true" by doing self-fertilization experiments. He realized that this would be an essential control to compare the appearance of different traits after future cross-fertilization experiments. There was a long section in his paper describing the experiments and the various features of the different plant hybrids that occurred after a succession of crosses. These included differences in pea type, whether round or wrinkled; differences in pea coat, whether white or gray; in shape of the ripe pea pods; in the position of the flowers on the stem; and differences in the length of the stems.

The next section gave the results for the first generation of plant hybrids bred from the initial parent plants; then results for the second generation, and then subsequent generations up to the tenth were recorded. He found when he crossed tall plants with dwarf plants the next generation were always tall; but when he crossed these tall hybrids with each other the dwarf feature reappeared in the ratio of 1 dwarf to 3 tall in the progeny.

As an example of some of his published results for the second generation of hybrids, he found from 7,324 peas, 5,474 were smooth and round, and 1,850 were angular and wrinkled giving a ratio of 2.96:1. Again out of 8,023 peas, 6,022 were yellow and 2,001 peas were green giving a ratio of 3.01:1. He never found an exact ratio of 3:1, not unexpected due to sampling errors. Even tossing a coin ten times rarely give a 1:1 ratio for heads:tails. In addition, plants cannot produce 0.01 or 0.96 of a pea, so it has to be rounded up to a whole number. Mendel was one of the first scientists to see the need for large samples to get accurate results. With a small number of plants, fluctuations can occur, and the actual ratio of 3:1 can only be found from the average of as many single experiments as possible; the greater the number, the more are chance effects likely to be eliminated.

A detailed analysis of the frequencies observed for other different inherited traits from generation to generation confirmed these results. But he presented a difficult algebraic model to show why this ratio of 3:1 appeared so consistent. This might have put Darwin off from reading any more of the article to find out where this ratio came from. Darwin said that *Mathematics in biology was like a scalpel in a carpenter's shop—there was no use for it*. Mendel's model also attempted to show which characters would reside in the pollen cells and which in the egg cells to explain the regularities in the appearance of this ratio of 3:1.

The final section made far reaching claims that the author had discovered laws of heredity that could predict the appearance of the different hybrid characters in successive generations of the edible pea, and that this would probably apply to other plant species as well. (At this point the meaning of a natural law in this context should be clarified: it is clearly not a divine command, nor is it an agreement among scientists that this should be so, but is more a principle or rule of nature validated by numerous replicated and coherent sets of experiments. Many biologists dislike using the word "Law" for living systems above the macromolecular level; they prefer the use of "Rule" or "Principle"). Of course, Mendel's results needed to be confirmed by further experimentation in different plants, but he suspected it to be correct in view of the unity in the developmental plan of all plant life. The final two paragraphs argued: (1) that the transference of features such as tallness and shortness among cultivated plants, such as the edible pea, seems to occur in discrete integral steps, and (2) if all of the steps accumulate in one species of plant this could definitely "transform" it into a different species. Mendel's conclusions left no place for Darwin's idea of blending inheritance that would imply the existence of a multiplicity of different gemmules coming from all parts of the plant.

Even by today's standards this is a "heavy" paper to read; but here were data and a model showing how the accumulation of small inherited steps in the structure of a plant species could become established in a population. And this could gradually lead by successive changes (due to Darwin's theory of natural selection) to the origin of a new species.

It is difficult for us to fathom Mendel's motivation for doing these experiments. It is clearly not for the fame of becoming a great professor at one of the major European universities; or the honor of being elected as a foreign member to a famous learned society, such as the Royal Society in London; or to amass a fortune from the sales of his publications (his article alone fetched $86,500 in 2013 at auction in New York; whereas Charles Darwin's *Origin of Species* First Edition sold for $60,000 in 2009— an example of relative values).

By 1865, Mendel had long taken the vows of a monk and priest in the monastery of St. Thomas. The monastery was more like a college run on the Augustinian maxim of *per scientiam ad sapientiam*: from knowledge to wisdom. The Augustinians were much less strict in their code of conduct than the other religious orders of the Catholic Church, such as the Benedictines or Carthusians. They valued teaching and research as well as Bible studies, fasting, and prayer. The main library was one of the most elegant rooms in the monastery; the floor was of polished parquet wood, and hardwood bookshelves lined three sides of the enormous space housing many of their book collection of around 20,000 volumes; the others were in an alternative study library where the monks did their daily bookwork.

The few dozen monks from St. Thomas, including Mendel, provided many of the teachers for Brünn's schools and philosophical institutes, including monks specializing in German literature, mathematics, and music. The Augustinian monks helped to found the National Science Society of Brünn. In 1865, Leos Janacek (1854–1928), the future composer, enrolled in the Abbey of St. Thomas to sing in their choir and play the organ adding further distinction to the place.

Gregor Mendel fitted into this Augustinian community very well except for having trouble learning the Czech language; his mother tongue was German. Naturally, he had to comply with the rules of the monastic community to remain chaste and celibate and to own very little personal property. Similar to most Augustinians he was not excluded from local society but was excused from doing local parish duties as a monk or priest. Due to innate sensitivity he could not bear to visit a sickbed or to see anyone ill or in pain, because this made him feel ill himself. His usual daily routine would be to arise early from a simple cell and attend two church services, morning and evening in St Mary's Basilica, an imposing Gothic building with an ornate clock tower that still chimes today. The rest of the day would be divided into Bible study at fixed times, and work in the monastery gardens or attending the Brünn Theological College for his training as a priest. He would have classes in Canon law, moral theology, Hebrew, Greek, Latin, and archeology. He became quite proficient in these subjects qualifying as a priest at a very young age.

He would have two cooked meals a day with much of the fruits and vegetables provided by the monastery gardens. Part of Mendel's work was to be one of the monastery's gardeners providing vegetables for the monks. However, his scientific nature soon prevailed and the monastery gardens became his research laboratory. He was given a strip of the garden of about 50 yards along the library wall of the monastery to do his experimental botany. The monastery abbot, Cyrill Napp, favored Mendel's work as he was an expert himself in the field of plant hybridization. Later, Abbot Napp had a large greenhouse built across the width of the lawn for Mendel's special use to extend his studies. Napp understood Mendel's plea that the greater number of plants studied, the more reliable the results will be. In Mendel's paper of 1866, he wrote that large numbers of plants are necessary because with small samples very considerable fluctuations may occur in the ratio. To establish true numerical ratios requires the greatest possible number of individual experiments to work with.

Mendel was ordained as a priest on August 6, 1847, just 15 days after he turned 25, the minimum age for this. He did not do parish duties but instead was given teaching duties. He rarely put on a priest's garb, but wore a long black cassock for the town. He taught mathematics, natural sciences, and Greek to boys in the third and fourth forms of the Realschule (an elementary school) in Brünn. His first big test came when he tried to obtain his teacher's certificate from the University of Vienna. He failed abysmally in the written papers to provide answers that any schoolchild would have known. The examiners passed him in the *viva voce* exam noting that he showed *no special brilliance or acumen* but that he showed *unmistakable good will and was devoid neither of industry nor talent*. Ironically, his lowest marks were scored in biology and geology. He failed the examination, although the examiners admitted that Mendel's biggest problem was that he was largely self-taught. If he were given the opportunity for more extensive study together with access to better sources of information then he would probably pass muster as an accredited teacher. With this in mind, Abbot Cyrill Napp granted Mendel a great opportunity.

The monastery would pay for him to attend the Imperial University of Vienna to study and improve himself.

From 1851 to 1853, Mendel had a glorious time in Vienna that transformed him from a self-educated Silesian peasant to a fully fledged natural scientist. He studied physics, mathematics, and the natural sciences, particularly botany. He became an assistant demonstrator at the Physics Institute, under the aegis of Doppler (of the Doppler effect fame), a position reserved for the brightest and best students. He learnt about the latest work on plant hybridization from Professor Franz Unger; and he was taught mathematical theorems involving combinations and permutations by Professor Ettingshausen. This was to stand him in good stead for the algebraic analysis of his own results. He returned to the monastery in Brünn in July 1853 and by the following year was starting to hybridize peas with enthusiasm. Unfortunately none of his notes or experimental notebooks have survived—most of his personal papers were burned at the time of his death to make room for the personal effects of the incoming abbot. One page exists of his handwritten results of his botanical experiments on the back of a piece of paper bearing the notes for a future sermon. Was his faith more important to him than his science?

He stated in his major published paper of 1866 that he was looking for laws governing the transmission of inherited characteristics of hybrid plants. He asked the right questions; and unlike so many scientists he obtained an exceptionally interesting answer. There appeared to be a constant ratio of 3:1 in the appearance of inherited features of the garden pea in some generations (namely the second). The way a scientific question is formulated is most important because this determines the design and analysis of the subsequent experiments. Asking the question: whether there are any mathematical regularities in the appearance of plant characteristics after crossbreeding two different varieties of the edible pea, as posed by Mendel, leads to a different set of experiments than taking up the question posed by Darwin and Galton: whether there are multiple hereditary factors coming from all the plant (or animal) organs that determine the inheritance of different characteristics in the progeny?

From 1854 to 1863, Mendel bred garden peas with only one major interruption when he returned to Vienna again in 1856 to try to win his teacher's certificate. This time he failed in the *viva voce* examination; so he left Vienna having to be content with remaining a nonaccredited teacher for the rest of his life.

Darwin had yet other chances to read of Mendel's work in the early 1870s. Hermann Hoffman, a professor of botany at Giessen, had written a small book on plant hybrids in 1869 and written on page 52 was a long excerpt from Mendel's paper of 1866. On Darwin's copy of the book (now preserved in the Cambridge University Library) are handwritten notes in the margins by Darwin on pages 50, 51, 53 (facing page 52), 54, and 55.

These are very close to the citation of Mendel's paper but it may be that Darwin skipped over page 52 or did not appreciate its significance.

Darwin had a further chance to read about Mendel's work in 1881. George Romanes (1848–1894, a student of Darwin's) was then preparing an article for the *Encyclopedia Britannica* on plant hybridization. He enlisted Darwin's help to suggest names of eminent botanists who should be included. Darwin replied by sending Romanes a copy of a recently received book by Wilhelm Focke on the topic, published in 1881. Mendel's work was summarized on three pages (pp. 108–111) and the section ended stating that: *Mendel thought he had found constant numerical relationships between the different types of crosses.* These pages were uncut in Darwin's copy and Romanes left them so. Mendel's name was included by Romanes in his article for the encyclopedia, but he never read what Mendel had done.

How did Darwin come to miss the significance of this one jewel out of the many plant hybrid papers that were being published at the time? On both occasions, he could have easily checked the results for himself. Darwin had personally done and was still doing large numbers of plant-breeding experiments using more than fifty plant species, including the edible pea, orchid, snapdragon, flax, primrose, and so on, but never with the idea of primarily studying the transmission of plant characters between generations. He was more interested in the problem of hybrid vigor and its role for evolution. His main question was whether seeds from cross-fertilized flowers would produce superior plants than seeds derived from self-fertilized flowers? It seems he never thought of performing plant-breeding experiments to check the results of Mendel, even though he had the required skill, knowledge, resources, and the patience to do this sort of work. In his book on *The Variation of Animals and Plants under Domestication* (1868) he wrote that he had planted 41 varieties of English and French edible pea to study the extent of their variation. Later, he published a book on further studies titled *Different Forms of Flowers on Plants of the Same Species* (published John Murray, London, 1877). He observed the variations that Mendel had studied: smooth versus wrinkled peas; tall plants versus short ones; differences in flower color or size, and so on; but he did not study any hybrids in detail. He did crosses using the common snapdragon (*Antirrhinum majus,* Figure 5.5) with the rarer (peloric) form (Figure 5.6). He was working with pure breeding peloric plants and crossing them with pure breeding normal type plants (differing in the size and symmetry of the flower of each type). In the second generation of hybrids that he obtained, he counted 90 to be the normal variety (with two as an intermediate type) and 37 to be the rarer peloric form, giving a ratio of 1:2.4. With the numbers of plants involved this is not statistically different from a Mendelian ratio of 1:3.[18]

Antirrhinum majus flore pleno

Figure 5.5 Snapdragon, *Antirrhinum majus*. The common blossom structure is shown, which Darwin (1868) crossed with the peloric variant, see Figure 5.6. All the progeny appeared in the common form, which when self-fertilized, yielded offspring in a ratio of common (*n* = 88) to peloric (*n* = 37) blossom structure of ~3:1. This was close to Mendel's results. (From Opitz, J.M. et al., *Ital. J. Pediatr.*, 42, 35, 2016. With permission.)

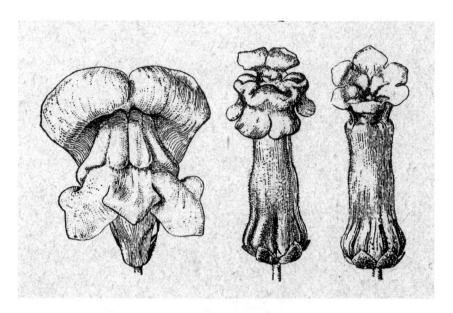

Figure 5.6 Common form (left image) and peloric forms (two right images) of the snapdragon blossom. The peloric form is fertile and hereditary. Charles Darwin explored crosses between the two forms of *Antirrhinum majus* (snapdragon) while researching the inheritance of floral characteristics for his *The Variation of Animals and Plants under Domestication* (1868). His results were largely in line with Mendelian theory. (From Opitz, J.M. et al., *Ital. J. Pediatr.*, 42, 35, 2016. With permission.)

He found a 1:3 ratio in his experiments with the evening primrose (*Oenothera biennis*, Figure 5.7) in the second generation of hybrids; he counted 75 of the common form of flower with a short stylus and 25 of the rare form of flower with a long stylus. However, he made no comments on these ratios. Darwin did an enormous amount of work counting seeds, weighing, and measuring them; planting them and measuring growth rates and the general hardiness of the progeny. However, he was still thinking along the lines of his gemmule hypothesis, so everything he measured had to be a continuous variable in which one would never expect to get constant reproducible ratios in the inheritance of any parental traits. It would all depend on the variable number of gemmules being shed from each part of the plant. So, he ignored the significance of his 1:3 ratios and made no comment on them. If only (a desperate cry made by many scientists when they realize they have just missed making a crucial observation) he had read and assimilated Mendel's article.

Some historians claim that Darwin did not even cut the pages of Mendel's original article to open it. He was not the only one to overlook the importance of the work. Of the forty reprints of Mendel's article,

Figure 5.7 Evening primrose (*Oenothera biennis*), the common form known as *pins*; and rare peloric form called *thrums*. Darwin studied crosses between *pins* and *thrums* and found ratios supporting Mendel's results but did not recognize this. (From Thome's catalog Flora von Deutschland, Osterreich 1885.)

one was sent to each of the following scientists: von Marilaun, Beijerinck, Boveri, Schleiden, and von Nageli; more calf-bound copies of the journal were to be found in learned societies around Europe, including the Royal Society, the Linnaean Society, and the Greenwich Observatory in Britain. No one really followed the results up until 35 years later at the turn of the century. Perhaps Darwin was put off by the obscure source of the article. An unknown Catholic monk working as an amateur alone in a remote country monastery, with no colleagues to check or discuss the data; and then publishing it in the journal of a local science society in an obscure

Austrian town, Brünn; perhaps all this gives one less confidence in the value of the work than if it had come from a major botanical center such as Paris, Berlin, or Vienna. It would also be unclear what was the monk's motivation for doing these studies; as mentioned before certainly not for worldly fame. Perhaps his motives were based on his faith and religion. John Milton, the poet (1606–1674) considered reading the book of God in the natural world is an act of worship in itself. It appeared that Mendel was doing pure science untouched by worldly motives.

The fact that a monk, a Catholic one at that, should even try to investigate the mechanism of heredity perhaps seemed ludicrous to Darwin. Many other scientists later on were to feel uncomfortable with the thought of an ordained Catholic priest achieving fundamental results in the field of heredity working by himself in a monastery (see the later attacks on Mendel by professors de Vries and R. A. Fisher). Darwin might take Mendel as just the gardener cultivating peas for the refectory and not doing proper science. However, Darwin always stored facts away in his mind however preposterous they might seem; this one he might have stocked right at the back in view of the later advice he gave his cousin Galton to do heredity work on peas. It was self-evident to him that blending inheritance does occur—you can see it in the skin color of children from mixed marriages. Yet "seeing" is not necessarily a "proof." One can see the Sun rise in the east and traverse the heavens to set in the west, but the Sun is not actually moving like this. If, like Galileo, one points a telescope at the planet Venus, one can see that it passes through a regular series of lightened phases such as the Moon. However, measurements of the positions of these phases show that this planet must be revolving around the Sun and not the Earth. Of course, we can close our minds to the contradictory evidence and just go on with our old geocentric beliefs. However, in the end, this will lead to navigational disasters.

The unsolved mystery therefore remains[19]: Did Darwin actually bother to read Mendel's article? A catalog of Darwin's library from Down House published in 1908 (that is 26 years after Darwin's death) did not record any of Mendel's papers. However, after Darwin's death in 1882, his scientific library passed to his son Francis. The Down House was cleared of its contents in 1896 following the death of Emma Darwin and the house then leased to a school. Francis Darwin later bequeathed the library to the professor of botany at Cambridge University and a catalog of the library was prepared by H.W. Rutherford (the one published in 1908). There was thus ample time for small items to go astray. However, the catalog did record the presence of both Focke's and Hoffmann's books; and the former undoubtedly mentions Mendel's claim to have found "constant numerical relationships" among the different structural forms of plants in the second hybrid generation of the edible pea.

Mendel aged 40 years had his first opportunity to travel abroad to London in 1862. He was selected to form part of the official Brünn delegation of about 30 strong to attend the first annual international exhibition a decade after the great *Crystal Palace Exhibition* of 1851. The present occasion was to display the new developments in technology and Mendel's job was to help to set up a display of crystal structures on the Realschule stand. Brünn was a center of industrial activity, mainly in textile manufacture and the town fathers believed the best way to develop the town's economy was to invest in the latest technical inventions. One of their ideas was to build a new technological museum in Brünn and the London Exhibition might be something for them to copy. The visit took about 3 weeks and Mendel had the chance to see many cities *en route*, including Salzburg, Munich, Stuttgart, Strasbourg, and Paris.

There is no evidence that Mendel ever tried to contact Darwin during his visit to London. Darwin by then was a celebrity and visitors would call at Down House unannounced to discuss evolutionary matters. However, at the time Darwin's twelve-year-old son, Leonard was seriously ill with scarlet fever and his parents stayed with him away from their house at Downe, receiving no visitors.

What would have happened if Darwin and Mendel had met in 1862? Probably Darwin would have been incredulous at Mendel's claims because according to the gemmule hypothesis you could never get exact mathematical relationships occurring from one generation to the next because of the variable number of gemmules that would be involved. He was also getting used to foreigners turning up at his house with far-fetched or even crazy ideas of their own about evolution. In the early 1860s, he was still very keen on the gemmule hypothesis, and so far there was no experimental evidence (apart for Mendel's as yet unpublished work) to refute it. He probably would not be diverted from his ideas after a short conversation with a Catholic monk; especially as the dignitaries of the Catholic Church, including the Pope, had been so harsh about him after the publication of *Origin of Species* (see Epilogue 1).

Darwin's mental lapse might have held up the progress of genetics in Britain for Darwin's students, especially for George Romanes, Thomas Huxley, and Francis Galton by at least three decades; and we would not have had the future acrimonious debate between the Mendelians and Darwinians in the early years of the twentieth century (see Chapter 15) to try to reconcile the issue of blending versus discrete (or binomial) inheritance.

chapter six

Cousins diverge

Diversity, controversy, tolerance—in that
Citadel of learning we have a fort
that ought to armour us well.

Blessed is the Man. M. Moore (1887–1972)

After the discussion with Darwin, Galton commenced his inquiry into the nature of inheritance of coat color in rabbits. He no doubt felt excited and honored to be working together with his famous cousin on such an important project. The connection would be bound to enhance his career.

He had no trouble obtaining the rabbits, but removing and transfusing blood was going to be more of a problem. His medical studies had made him familiar with cannulating blood vessels; after all bloodletting was one of the simplest operations he had learned in Birmingham. He had not practiced the procedure for so long that at first he thought he would probably kill the rabbits rather than transfuse them.

He started with three white bucks and four white does, which were bled and then transfused with blood from the common black, lop-eared rabbit. He initially worked up the number of transfused rabbits to 32. The experiments were thorough and accidents rarely occurred. The apparatus was quite basic but served its purpose very satisfactorily. It was astonishing to see how quickly the rabbits recovered after the effects of chloroform anesthesia had worn off. Their spirits and mating instincts were in no way dashed by the blood transfusion, which only a few minutes before had in some cases replaced up to one half of the blood in their bodies. Galton's part was only to insert the cannulae and to collect the blood. He was lucky enough to recruit a Dr. Murie to do all the more delicate and difficult work. Dr. Murie was the prosecutor at the zoological gardens in London. When Galton called on him to discuss the project at the zoo and invite him to join in, he found a dead cobra lying on Dr. Murie's table. Galton told Dr. Murie that he had never properly seen a snake's poison, so the doctor very coolly opened the creature's mouth, pressed firmly at exactly the right spot, and out came that most delicate and wicked looking fang with a drop of venom exuding from it. Galton thought that a man who was so confident of his anatomical knowledge and had the nerves to attempt such a daring act must be a suitable person to deal with the gentle rabbit, which he agreed to do.

They began experiments at Galton's house in Rutland Gate. Galton built hutches in his backyard to accommodate them, but Louisa soon started to complain that her house was being turned into a rabbit warren. The experimenters had to transfer their activities to the physiological laboratories at University College London to obtain more tranquil conditions. The experiments took 2 years to complete. Even Darwin's wife became involved. She wrote a letter to her daughter Henrietta:

> *Father is wonderfully set up by London, but so absorbed about work etc. and all sorts of things that I shall soon force him off somewhere before long. Francis's experiments about rabbits (viz. injecting black rabbit's blood into white and vice versa) are failing which is a dreadful disappointment to them both. Francis said he was quite sick with anxiety till the rabbit's accouchements were over, and now one naughty creature ate up her infants. He wishes this experiment to be kept quite secret as he means to go on, and he thinks he shall be so laughed at....*

Various techniques of blood transfusion were tried and toward the end of the year they had established a method of cross-circulation of blood between the carotid arteries of the two different strains of rabbits by which as much as 50 percent of the total blood volume could be interchanged. Later next year Galton wrote to Darwin:

> *Good rabbit news! Two injected blacks have produced an infant marked with a white fore-foot. It was born April 23rd but as we do not disturb the young, the forefoot was not observed till to-day. The little things had huddled together showing their backs and heads, and the foot was never suspected. This result is from a transfusion of only 1/8th part of alien blood in each parent; now after many unsuccessful experiments I have greatly improved the method of operation and am beginning on the other youngsters of my stock. Yesterday I operated on two who are doing well today and who have a third alien blood in their veins. On Saturday I hope for still greater success, and shall go on at any waste of rabbit life until I get at least half alien blood.*

It seemed Galton was hoping against hope to prove Darwin's hypothesis to be true. However, it was one of the several false leads; the appearance of a white forefoot is a common event with these black rabbits and had nothing to do with blood transfusions producing mongrelization of the offspring.

Galton kept detailed records of all the experiments in his laboratory notebooks. He kept Darwin fully informed of all the experiments that seemed to corroborate Darwin's ideas; and eight of the rabbits were sent to Down House for Darwin to participate in the experiments. Galton gave details about these eight young rabbits, how they should be mated, and when the young should be returned to London for further blood transfusions. He was very particular that the bucks and does should be kept in separate hampers, and the name *bucks* was to be written on the correct container. They were delivered back by hand to the university laboratories for a Mr. Carter to receive. The rabbits arrived safely and were pretty lively after their journey and by next morning were well able to perform their stuff. After 2 years of such hard work Galton thought he had sufficient data to publish a paper.

The quarrel[20]

Galton went to visit Darwin at his house in Downe village to show him the final results. He usually arrived at Orpington at 11.12 a.m., and so could reach Downe by 12.30 p.m. The Darwins never owned a carriage; these were mostly run by the doctors in the neighborhood who drove about in broughams in their top hats and frock-coats. The Darwins initially had a donkey cart and sometimes two donkeys were driven in tandem. Recently they had upgraded to a pony and trap to bring visitors, usually relatives, from the station to their home. The pony-trap swept through the tunnel under the railway line heading for Downe village along country lanes.

After a further drive of about four miles along deep narrow lanes where the trees met overhead they came to the village and passed the simple old church before turning past the blacksmith's shop and the village pond to reach Down House. Galton's heart sank; he thought Down House a very dull and uncomfortable place. There was hardly any local society and the scenery was rather flat and uninteresting, mainly clay and chalk low hills. The food was always very simple and not at all to Galton's taste: lunch was usually shepherd's pie and rich creamy brown rice pudding with prunes; tea was often sponge cake with honeycomb.

A great inconvenience of the house was the lack of a bathroom, or any hot water except in the kitchen. There were plenty of housemaids to fetch and carry hot water in brown painted bath-cans to the basins in your room standing at the foot of the great four-poster beds. The downstairs rooms were all large and furnished with dignified and plain furniture reflecting the barer way of life of the earlier part of the century. The menservants room was a long dark attic with a board floor with three beds and hardly anything else. Here the coachman, the butler, and the footman all slept together in very barren quarters. Harriet was the head housemaid. She had a rich voice, lovely laugh, a strong Kentish accent and was beautiful to look at.

Darwin was glad to greet his young cousin on the doorstep to the rambling old house. Galton followed him into the house, took off his overcoat and hat and they went straight away into Darwin's study, which is the quietest place in the house away from the noise and bustle of the numerous children. They could go through Galton's results in peace.

The study was a rather poky, untidy room on the ground floor near to the stairwell. It was lined with miniature cabinets where Darwin stored most of his files and paper records, mainly a leftover habit from his time of study onboard the Beagle where space was at a premium. There were piles of books lying around that needed to be read, many on loan from the London Library and delivered by post. To many visitors surprise there was only a simple dissecting microscope on the central table, with a few old-fashioned feather quill pens for making notes. Darwin preferred this to the compound microscope that most other biologists were using at the time. The walls to the left of the fireplace were screened off and Darwin took his water therapy behind the screens if his health was in bad shape. The room was out of bounds for the children unless they had a good reason for entering. They were scared of one of his walking sticks that he kept there. It was a very slender whalebone cane topped with an evil-looking skull, which the children believed was a shrunken head from some savage tribe; in fact, it was just a human skull carved in ivory.

Darwin settled himself in his large mobile armchair. This armchair had been specially built for him because he disliked writing at a desk; he preferred to sit in this chair with a board across the arms to support his writing materials. His children enjoyed scooting and swiveling themselves around in the chair whenever they were allowed.

Galton handed over the laboratory notebooks for Darwin to inspect. He and Dr. Murie had worked solidly for 2 years and their results came to this. They had the results of 124 cross-transfused rabbits from 21 litters and none of the offspring showed any signs of a change in fur color after transfusion of foreign blood into the mother. As an explanation for the negative results, Darwin wondered whether the gemmules from one rabbit did not survive for long enough after transfusing into the other rabbit for accumulation in the ovaries or testes to occur. Perhaps they may only survive for a couple of hours. However, Galton had thought of that and the rabbits on the marked pages were from matings made within one hour of the blood transfusion. As Darwin could see there were no signs of fur changes in their litters either.

Another notebook contained all the results for the second generation whose parents were the offspring of the original cross-transfused rabbits. Again one could see that the results were negative there too. Darwin thought there was still something wrong with the experiments. Perhaps the design was wrong. Galton had sent Darwin at least five letters during these experiments asking Darwin's advice about the procedure and Darwin had replied to them all approvingly.

Still the design could be wrong. Galton should examine Darwin's arguments more carefully in his book on *Variation of Plants and Animals under Domestication.* Had Darwin ever written that gemmules are to be found in the blood? Darwin wanted to be precise about what he wrote. The gemmules do not need to circulate in the blood. The idea works just as well for plants that do not have a blood circulation. Darwin wrote that gemmules might circulate in the body fluids. He never mentioned blood.[21] Galton was perhaps generalizing too much from just one type of experiment. Galton thought that this was a quibble. How can one expect the gemmules to get to the sex organs, from say the brain, if they do not pass through the blood stream?

Darwin went on that there was nothing inconsistent between his hypothesis and Galton's results. He had worked before on the data that were negative but could still lead to ideas that were adequate for certain purposes. Darwin considered that the correct conclusions of Galton's experiments were quite simple. After a limited number of experiments in rabbits he had not been able to find any evidence that gemmules in the blood affected inherited characters such as fur color—that was all. It was not a decisive *No* to Darwin's ideas—but even possibly an inaudible *Yes* if they would only do more extensive experimental work.

This was being rather hard on Galton. If Darwin had believed that why on earth did he agree in the first place for Galton to do the blood transfusions? Darwin had encouraged Galton to go on with the work and gave plenty of technical advice on the experiments in his letters. And now two years later he tells Galton that the gemmules are not necessarily to be found in the blood.

Galton perhaps thought that Darwin was just trying to ignore the evidence to save his hypothesis. Anyway, Galton was going to publish the results of his work. And since Darwin was there at the start of the project and had helped to set it up Galton hoped that Darwin would agree to be a coauthor. Darwin may have considered this but decided that he would not come on as an author. He recommended Galton not to publish such incomplete results that showed nothing. It would without any real evidence cast doubts on his ideas and just confuse the public perception of them.

But if the hypothesis could not be verified?

Darwin considered that Galton had not shown this. He had not done enough work to prove it untrue. The investigation was not such a simple matter. One needed to use other experimental models. One should repeat the transfusion experiments using different body fluids such as lymph where perhaps the gemmules are really to be found. One should try different living systems—perhaps even plants.

However, at the time Galton was still keen to publish his findings even without Darwin as coauthor. Darwin remained adamant. He had scored a marvelous public success with his theory of natural selection. It would be distasteful for him to be shown wrong at this late stage in his career on such an important topic of heredity. His views were perhaps those of an over cautious and thoughtful scientist. If one thinks that one's hypothesis is valid and the first experiments fail to support it, it is not unreasonable to infer that one's experimental design or even the methods that have been used are inadequate. One must think of other ways of testing the hypothesis using other experimental systems. He believed his concept of gemmules could turn out to be as important as Newton's idea of universal gravitation. You cannot see or touch either but both have tremendous explanatory value. Gravity explains the motion and orbits of the planets; gemmule explains the whole field of heredity and must prove as important to biology as gravitation has been for physics. Obviously Darwin could not stop Galton from publishing the results. He could still point out publicly all the flaws in Galton's experiments. As Darwin's reputation as a scientist was much greater than Galton's, he hoped that he could dissuade his young cousin from publishing and that his hypothesis would survive.

It was a decidedly difficult position for Galton to be in; it might appear that Darwin was less interested in the experimental work than in his desire to maintain his ideas in spite of any evidence to the contrary. Darwin had found self-effacement difficult with the preemptive paper by Alfred Wallace on natural selection and it might be so again with his ideas on inheritance.

Galton thought he should publish; although Darwin was the acknowledged authority and now one of the most famous naturalists in the world, he might not necessarily be correct in every idea that he had.

Darwin thought that any scientific cause would not be helped by publishing the results of "half-baked" experiments that were uniformly negative. It would just confuse public opinion about his work and this may even spill over to his theory of evolution. Possibly Darwin was becoming too easily upset by public criticism. His emotional reserves that allow one to cope with such public disputes had certainly diminished with the passing years. He admitted in older age that he could no longer enjoy music, although his wife was a competent pianist; that he could not bear to read poetry; and that even Shakespeare was so intolerably dull that it nauseated him to have to read his work. However, he could still concentrate on the masses of facts he had collected about natural history. Studying and drawing the sexual organs of thousands of barnacles for 9 years for one of his earlier books was what he was really good at. No wonder Darwin's small son asked a friend: "Where does your father do his barnacles?" as though that is all that fathers could ever do.

On the way back on the train to London, Galton may have thought of the phrase from that old grammarian, Apollodorus: "We need new heroes..." Darwin was of course Galton's hero, but perhaps every hero becomes a disappointment in the end. We attribute qualities to them that we wished we had ourselves only to find them lacking there too; they have human frailties similar to the rest of us. Yet he had to admit the excellence and thoroughness of Darwin's works on evolution; and his other books on barnacles, coral reefs, and the descent of man.

Then there were all the honors that Darwin reaped for his previous publications. He had been awarded the *Merit of Honor* from Germany, he was being put up for an honorary degree of LL.D. from Cambridge University, and a marble bust had been commissioned of him to be sculpted by T. Woolner R.A. Then Darwin's health was not good. It was hitting below the belt to denigrate Darwin of his loss of feeling for music and poetry, which may be due to his prolonged illness. Then again other people said this illness was just an excuse to stay at his house in Downe and let all his friends do the fighting on behalf of his ideas to the world outside. It could be pure cowardice on Darwin's part if all these head-aches and indigestion were just pretence to keep out of the fray of scientific debate about gemmules.

Perhaps it was best to forget about personal relations and ambivalent feelings about his senior colleague. If he would just stick purely to the science, he felt he really could not go wrong. Galton had done two years of hard work on his project as honestly as he could. He felt it was the best he could do. Galton decided to publish the results whatever Darwin wanted. Just because Darwin was right about natural selection, it does not mean to say he was going to be right about every other hypothesis he held. These perhaps were Galton's thoughts that determined him to go ahead with the publication come what may. He knew at least one person who would read the article with care and interest: Cousin Charles.

A few months later Darwin received a pile of journals by post, which he hurriedly took to his study to scan their contents. Sure enough there was an article by Galton in the Proceedings of the Royal Society (1871) titled *Experiments in Pangenesis dealing with the fur color of rabbits after blood transfusions*. Its gist was exactly as Galton had told him a few months before at Downe and it started with these challenging words: *I have now made experiments of transfusion and cross-circulation on a large scale in rabbits and have arrived at definite results negativing, in my opinion beyond all doubt the truth of the doctrine of gemmules*.[22]

His young cousin had gone directly against his wishes and seemed to be spoiling for a public quarrel. He had dashed off this paper saying that the theory was proved untrue, even after having consulted Darwin's opinion. It would be very unusual and against all scientific etiquette, for a junior author to send off a paper for publication against the expressed agreement of a senior participating colleague even if he did not want to appear on the article himself as an author. At least there was nothing more to be disclosed behind his back, as the publication just repeated what Darwin already knew from Galton's last visit to Downe.

Darwin had already taken a lot of criticism about his gemmule hypothesis and was by now exceedingly sensitive to adverse comments. He was very quick to stand up for his hypothesis. A month later, a defense was published in *Nature* 1871, volume 3, pages 502–503 and titled *Pangenesis*:

> *In my writings on inheritance I have not said one word about the blood or about any fluid proper to any circulating system. It is indeed obvious that the presence of gemmules in the blood can form no necessary part of my hypothesis, for I refer in illustration of it to lower animals, such as Protozoa, which do not possess blood or any vessels; and I refer to plants in which the fluid, when present in the vessels cannot be considered as true blood. Mr. Galton concludes from the fact that rabbits of one variety with a large proportion of the blood of another variety in their veins do not produce mongrelized offspring, that the hypothesis of gemmule transmission is false. It seems to me that his conclusion is a little hasty. As it is I think everyone will admit that his experiments are extremely curious and that he deserves the highest credit for his ingenuity and perseverance. But it does not appear to me that the gemmule hypothesis has been proved to be untrue.*

Galton became quite upset when he read this and tried to patch up their animosities immediately with a further hurried letter to Darwin giving recognition to his superior status and authority: *I am grieved beyond measure to learn that I have misrepresented your doctrine and the only consolation I can feel is that your letter to Nature may place that doctrine in a clearer light and attract more attention to it. You will see my reply to your letter in next week's Nature. I will justify my misunderstanding as well as I can, and I think reasonably. I will begin an entirely new and different set of experiments tomorrow unrelated to inheritance. Very sincerely yours Francis Galton.*

The older man had succeeded in scaring off Galton on the whole topic of heredity for the time being.

Some days later, Galton went to visit his favorite sister Adele in Launceston, Cornwall to seek her advice about the draft of his letter to *Nature* in reply to Darwin's letter. He relied on her opinion and good sense in times like this. He was told quite simply that he could not quarrel with his cousin in public like this. The letter was downright rude. Darwin was a famous man. The world respected his opinions. He is part of the family. And Galton used to admire him so much in the past. Adele could not understand why Darwin needed to be treated like this. And all about something called gemmules of which she knew, and needed to know nothing, and cared even less. It was absurd to let such a silly fuss and bother about trifles grow to so immense an issue.

For Galton it mattered very much. He thought the subject extremely important and cared about it more than anything else, and his first draft appears to have been too aggressive. Adele tried to soften the impact of her brother's letter. Galton had to agree. He had respected and trusted her judgment from his earliest childhood. Some paragraphs were to be amended, and it was published verbatim in *Nature* on April 27, 1871:

> *I do not much complain of having been sent on a false quest by ambiguous language for I know how conscientious Mr. Darwin is in all he writes, how difficult it is to put thoughts into accurate speech, and again, how words have conveyed false impressions on the simplest matters from the earliest times. Nay, even in the idyllic scene which Mr. Darwin has sketched of the first invention of language, awkward blunders must of necessity have often occurred. I refer to the passage in which he supposes some unusually wise ape-like animal to have first thought of imitating the growl of a beast of prey, so as to indicate to his fellow monkeys the nature of the expected danger. For my part, I feel as if I had been assisting at such a scene. As if, having heard my trusted leader utter a cry, not particularly well-articulated, but to my ears more like that of a hyena than of any other animal, and seeing none of my companions stir a step, I had, like a loyal member of the flock dashed down a path of which I had happily caught sight, into the plain below, followed by approving nods and friendly grunts of my wise and most respected chief. And now I feel, after returning from my hard expedition, full of information that the suspected danger was a mistake, for there were no signs of a hyena anywhere in the neighborhood. I am given to understand for the first time that my leader's cry had no reference to a hyena down in the plain, but to a leopard somewhere up in the trees; his throat had been a little out of order—that was all. Well my labour has not been in vain; it is something to have established the fact that there are no hyenas in the plain, and I think I see my way to a good position for a look out for leopards among the branches of the trees.*

When Darwin read this satirical piece in *Nature* a few weeks later, he was not particularly amused to be compared to an ape-like animal. There was to be a very strange ending to this episode. Darwin roped his cousin back into the transfusion project and they still continued to do more rabbit experiments together for another 18 months; perhaps under gentle persuasion by Darwin, and to give time for Galton's phase of rebellion to pass. The results were always negative. After some time they became discouraged and decided to give the problem to someone else. It would have been a better science to choose an entirely different set of experiments to test their hypothesis, but it is surprisingly difficult to change tactics after one has gone a certain distance along a particular research path. It was only much later that they thought of doing experiments in plants rather than animals.

Meanwhile, Mendel had quietly cultivated his peas apparently without any reported fuss or bother, analyzed his data, and written his paper. It was first read before a meeting of the Natural Science Society of Brünn on February 8 and March 8, 1865. The audience of about forty probably received it politely, because Mendel was a well-liked figure in the community, although they were perplexed. They were mainly agriculturalists, botanists, chemists, and doctors. No one asked a question probably because no one understood what Mendel had discovered or realized its implications. The lecture was reported in Brünn's daily newspaper, the *Tagesbote*. Josef Auspitz was on its editorial board and was also Mendel's superior at the Realschule as the headmaster. A creditable account of the lecture was produced, perhaps written by Herr Auspitz himself: *the numerical data with regard to the occurrence of the differentiating characters in the hybrids and their relation to the stem species were worthy of consideration.*

Mendel published his work in 1866 titled *Versuch uber Pflanzen-Hybriden* (Experiments in Plant Hybridization) in the Verh. Naturf. Ver. in Brünn. It was customary with lectures delivered at its monthly meetings for the Brünn Society to publish completed papers in its official proceedings. Mendel requested forty reprints from the journal editor, an excessive number for that time, and set about publicizing his results as best he could. We know for certain the fate of twelve that he sent out and also know that they were promptly ignored by the rest of the scientific community apart from Carl von Nageli working in Munich. He opened a correspondence with Mendel trying to discourage Mendel from further working along these lines. Von Nageli had developed his own theory of inheritance involving a part of the cytoplasm that he called the idioplasm. The idioplasm had some marvelous properties of being able to transmit inherited characteristics. He clearly did not like rival theories and suggested a project with the hawkweed (*Hieracium*) for Mendel to pursue. This could not possibly work for Mendel's type of experiments because the plant mainly reproduces asexually, by a sort of cloning. However, Mendel did experiments with another plant species, the runner bean (*Phaseolus coccineus*, Figure 6.1) that he described at the end of his 1866 paper. He found the same rules as applied to the edible pea but using much smaller numbers.

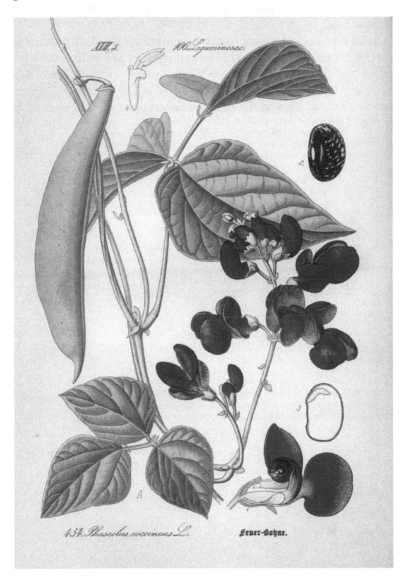

Figure 6.1 Mendel confirmed his results on peas in crosses with the bean, here *Phaseolus coccineus* again getting ratios of 3:1. (From Opitz, J.M. et al., *Ital. J. Pediatr.*, 42, 35, 2016. With permission.)

Von Nageli came to write up his ideas on the idioplasm and the nature of inheritance in his great work of 1884 titled *A Mechanico-Physiological Theory of Organic Evolution*. Although dealing with crosses between a variety of different plants and animals, there is not a single mention of Mendel's work despite their discussion of results that they had by correspondence over seven years. Another great opportunity to advance the subject was lost.

Mendel's paper was eventually reprinted in English translation in 1901 in the Journal of the Royal Horticultural Society after many people started to realize how groundbreaking his ideas were. William Bateson incorporated this translation into his book titled *Mendel's Principles of Heredity* in 1902, which was the first published book about genetics, a new word coined by Bateson.

chapter seven

The referee[23]

Truth—Be busy to seek her; believe me this,
He's not of none, nor worst, that seeks the best...
Doubt wisely; in strange way
To stand inquiring right, is not to stray;
To sleep or run wrong is.

On Religion (Satire111). John Donne (1572–1631)

George Romanes (1848–1894) by the age of 25 was a friend to both Darwin and Francis Galton and he was asked to act as a sort of referee between them. They wanted him to repeat some of the critical experiments to resolve the dispute about the rabbits. He came from Kingston, Ontario, where his father was a professor of Greek at Queens College. Despite his unusual name his father directly descended from an old Scottish family who had been settled in Berwickshire since 1586. In 1848, his father inherited a considerable fortune that allowed him to give up his Professorial chair at Kingston and he brought the whole family (three sons and two daughters) back to England. They settled in 18 Cornwall Terrace, Regent's Park, and George had a happy childhood until his schooling was interrupted by illness, a bad attack of typhoid. Afterward he went to Gonville and Caius College, Cambridge. He enjoyed his time there and was especially keen on boating but not particularly skilful at it. He recalled falling into the cam several times. He remembered looking calmly and intelligently at the green bubbles going slowly past him as he went down head first into the depths of the river. He once narrowly escaped drowning. There was a young woman in the same boat, Jane Mitchison, accompanied by her small brother as a chaperone. When she saw George fall in, Jane rose grandly to her feet with a ringing cry of "Oh George" and simply stepped into the river after him, seeming to prefer a watery grave herself than never to see George again. However, one gentleman was already swimming about in the stream and another dived splendidly into the river after him. So, George was happily rescued twice. Jane managed to clamber back into the boat by herself.

The highlight of his student days was to win the Burney Essay Prize of 1873; and he was surprised and gratified at the congratulatory notices he received. Another fortunate event occurred in 1873 when he published a short letter in *Nature* about some of his youthful views on evolution. It was not a particularly exciting letter and he was surprised that Mr. Darwin picked up on it despite its somewhat pedantic style. It concerned the coloration of flatfish such as the flounder or plaice. Is the coloration an example of reversion to an ancestral type as suggested by Mr. Saville Kent or due to evolution of a new species? To a self-professed Darwinian such as himself he published in the letter:

> *Nothing can be more evident that the coloring has been acquired because of its protective adaptation to their peculiar form and habits. But it is difficult to see how such coloring could have conferred protection upon their free-swimming ancestors, so that unless we make the highly anti-Darwinian supposition that the common progenitor was colored in anticipation of the habits and life-style to be adopted by its offspring there are only two hypotheses open to us, viz., that the unmodified progenitor adopted through natural selection, the habit of lying on its side because of its original sandy color ... or the other supposition, as being the most probable, that the coloration of flat-fish is the result and not the cause of their form and has therefore been acquired during the process of their flattening.*

The letter went on with some rather undergraduate explanations and he gave an outline of some easy experimental tests by undertaking a series of crosses between the normal and the piebald colored fish.

It was extraordinary that Mr. Darwin actually replied to it, by personally sending him a congratulatory letter. It was as though the Queen of England had written to him. Darwin later invited him to visit Downe. It began a new epoch in Romanes' life. He still remembered Mr. Darwin's first words to him: *How glad I am that you are so young!* It is all very well having a promising talent at 25; the difficulty is to retain it at 60. Yet it started with an unbroken friendship marked by reverence and affection on the part of Romanes, and by an almost fatherly kindness and interest in his career on Darwin's part who was now 64 years old.

Darwin adopted Romanes as a sort of protégée. The young man was very handsome (Figure 7.1) and had an appealing eagerness, directness of speech, and a habitual keen expression. Darwin liked his youthful enthusiasm. He was always looking out for a disciple to carry on the torch of his work and Romanes helped him considerably with some of his work on animal behavior. Romanes later wrote a book of his own about this titled *Animal Intelligence* that he had published in 1882, the year of Darwin's death. He made quite a career for himself mainly by association with Darwin.

Figure 7.1 George Romanes aged about 35 years.

Romanes first met Galton after reviewing one of his books in the Spectator. Romanes made it a good one: *Galton has no competitor in regard to the variety and versatility of his researches,* flattery but also true. Galton wrote to thank him afterwards:

> *I have just read the splendid review of my book in the Spectator. I was delighted by it, but at the same time you over estimate the value of what I do, and I do not feel worthy of such praise. I cannot think how you have so much time to spend on another's work when you have yourself so much in hand. I hope that your work is progressing well and perhaps we can meet later in the autumn.*

They arranged luncheon together after that on several occasions and became good friends. Romanes told Galton that he never allowed personal friendships to influence what he wrote in reviews; and if he was so uniformly biased as to overestimate the value of what Galton wrote, he knew that he shared this opinion with many other scientific men. At least he knew if he or anyone else had written the book, anyone else's judgment would have been the first to endorse all that he had written.

Therefore, Romanes was in a good position to understand and appreciate both the viewpoints of Darwin and Galton. He was only a junior scientist in the field, a foot soldier in the fight against ignorance and superstition. Young scientists are an essential breed and are much undervalued; they provide the bricks and mortar for men such as Darwin and Galton to build their airy edifices and grand unified theories, or to challenge the dogma of the generally accepted model.

Now a clash of doctrines (gemmule versus something else) is not a disaster—it is an opportunity. There is no better sign of maturity in a science than a crisis of principles. The clash is a sign that there are wider truths and finer perspectives to be found in a more subtle science. It indicates that the basis of the currently held science finds itself so sure and confident in its ruling principles that it can afford the luxury of submitting its newest theories boldly to revision, which means that it demands of them still greater vigor, robustness with no loose ends. The intellectual vigor of a man can be measured by the dose of skepticism and doubt that he can assimilate. Part of the difference that arose between Darwin and Galton was their totally different conceptions about how to practice science. Darwin was a great naturalist; he was a shrewd observer and had a vast curiosity about the natural world. He believed in experimentation but this was mainly based on simple but well thought out field experiments or horticultural trials, rather than on laboratory-based techniques. He would then try to group and arrange all the facts together to arrive at some unified theory that could explain the different circumstances that are found in nature. Once he had developed a good theory he held to it rather strongly: 2 + 2 always equals 4.

Galton on the other hand, was much more of an innovator and experimentalist. He could think of $2 + 2 = 5$ by just altering the meanings assigned to the symbols. It was all a question of the way you look at the terms. He was not just a passive observer of nature but he could multiply his hypotheses by asking different questions. He was always asking questions about *What if we tried this...?* Consider, for instance, the case of a man falling out of a window to his death. Galton could think of the incident in many different ways: as for the psychology it may be a case of unrequited love leading to a suicide jump; as for the physiology the loss of life may be due to a skull fracture damaging vital brain centers; as for the physics it is a case of a body impacting at high velocity on a harder body due to gravity, and so on.

Galton liked to test hypotheses by deducing their consequences and then seeing if they could be verified by experimentation. He especially liked experiments that produced numerical data on which he could do statistical analyses, which he loved. One of his favorite mottos was *Whenever you can, count it*. At certain points our ideas and hypotheses touch the reality of nature; and these points can lead to critical experiments. The experiment is a manipulation of natural events, to isolate and control the circumstances, and to try to force Mother Nature to respond to a question that we ask in a particular way. It requires ingenuity to think of new arrangements that will decide the issue. Moreover, this requires an up-to-date knowledge of the problem as well as the faculty of imagining new solutions. Galton tried to experiment and to measure almost everything, even the beauty of women. Few scientists have made such imaginative and lasting contributions to as many fields as Galton. He was in many ways an opposite type to Darwin who just preferred to concentrate on one or two major projects. Galton was one of the founding fathers of the science of meteorology, discovering and naming the anticyclone and one of the first to construct weather maps. (Mendel also studied meteorology. Since 1865, he was asked to become the chief weather watcher for the city of Brünn. For about 27 years he took regular meteorological readings from the barometer and thermometer hanging in the abbot's residential quarters at 7 a.m., 2 p.m., and 9 p.m. After accumulating readings for a month he would send them to the central weather station in Vienna.) Galton was a pioneer in the use of fingerprints to identify people. He initiated the field of modern psychology being the first to use twin studies to untangle the differences between nature and nurture in the formation of the character of people. He also used word association studies to explore a theory of unconscious thought processes well before Freud. He coined the word "eugenics" and started this movement in the latter half of the nineteenth century. He invented several statistical tests, including correlation analysis to see if a biological factor (such as the height of parents) relates to another (such as the height of their children), which is used to this day. Of course, such a relationship does not establish causality; it could be due to a confounding effect. The oft quoted example is a small village where the number of storks observed flying overhead correlates with the number of new infants born in the village; does this mean that storks are the bearers of these newborn infants? No—the confounding effect could be that as more children are born in the village the number of families and therefore the number of houses increase that provide more roofs for nesting places for storks. So, the increasing numbers of storks in proportion to babies are coincidental, only relating to the number of houses.

Therefore, when the dispute about gemmules had reached the public domain with angry letters published between the protagonists, Romanes was called in to try to establish the truth of the matter. Romanes had been present at many of the conversations on heredity, with either Galton or Darwin discussing them with him. He was trusted implicitly by both of them.

After visiting each of them in turn to discuss the matter, it seemed best if Romanes would try to repeat the experiments independently. This would save face for both Galton and Darwin. Romanes believed that Darwin really thought that gemmules might in fact circulate in the blood of animals. All Romanes had to do was to establish one of two opposite opinions (gemmules exist; gemmules are fictions) and to see which idea would lead to useless, and which to fruitful results. If the idea was any good it should lead to new questions, hence new studies, and then to new facts that fit with the body of already established knowledge. There were indeed some other very interesting points to study and Romanes felt he had the experimental skill to enlarge considerably the field of enquiry. In the end, he just decided to use exactly the same protocol as previously set down by Darwin and Galton and even used some of the same surgical instruments that Galton had used. He made several improvements in the methods with the kind help of a Professor Schafer. He chose wild rabbits to supply the blood and Himalayan rabbits to receive it because they are such a pure breed showing constant bodily features and fur coat color. Any foreign characteristics would be easily spotted. He employed the best method for cross-circulation between the carotid arteries of the two strains of rabbit, as perfected by Galton in the second year of his study. Instead of single cross-transfusions he did multiple transfusions into the same rabbit, so as to replace more of the blood. In some experiments, he even bled three wild rabbits completely to transfuse all of the blood into a Himalayan rabbit, so that almost all its blood was replaced. At the end of his studies, Romanes was duty bound to report that the experiments were again uniformly negative, just as Galton had described. Every great scientist before Darwin has made bold guesses, and there never was a bold man whose guesses were not sometimes wrong. The real test of greatness is if he can admit, simply and humbly, that he was wrong. An essential aspect of creativity is not being afraid to fail or to look ridiculous.

As Romanes had firmly come to the conclusion that Darwin was wrong that settled the matter for him. Darwin expressly wished that these later experiments should not be published and Romanes naturally complied with the old man. He had seen the difficulties that Galton had got into by disregarding Darwin's wishes and his own career so much depended on Darwin's good will. Darwin even went on to encourage Romanes right up to the year before his own death to undertake further experiments on the gemmule theory. The experiments were to transplant the seeds of one variety of plant into the ovary of a different species of plant to see what characteristics the next generation would take. Would the seeds of the host plant modify the features of the developing alien seeds by transmission of gemmules? Again these experiments were negative; there was no evidence for the presence or action of gemmules.

Darwin was delighted when a friend of his son, Mr. Balfour, a young biologist from Cambridge, came to visit Downe. *He is very clever and full of zeal for Nat. Hist.—He has been transplanting bits of skin between brown and white rats, in relation to my gemmule hypothesis* he wrote to Galton.

What amazed Romanes was something he never really understood about Galton's character. Galton still went on working for almost another 2 years on cross-transfusing rabbits and always, like him, with negative results. His short wave of rebellion against authority was over. It seemed that Galton's reverence and hero worship for Darwin returned in such a full force that his respect for Darwin's authority and judgment knew no bounds—even to the extent of working on ideas that he knew had no experimental support. It appears that almost all of Darwin's rabbit letters encouraging Galton's experiments have been lost or possibly destroyed by Galton later on. Only the letters that followed publication of Galton's paper on the hypothesis are preserved because the whole issue was now made public. This would have the effect of bolstering Darwin's reputation; and again Romanes wondered why Galton would want to do this having been so badly let down.

By late 1873, Galton had completely backed down from the quarrel. Once more he approached Darwin to enlist his help in a new project that he had just started. He was trying to define the mental attributes that define an eminent man of science. What does such a person need to be successful as a scientist? His method was to send questionnaires to all the fellows of the Royal Society to see if there were any common themes among them. The questionnaire was very detailed running to seven pages and asking about family background and accomplishments, health, personality characteristics, religious beliefs, education, and scientific achievements, and so on. He published the results in a book in 1874 on *English Men of Science, their Nature and Nurture*. Darwin did his best to help with the survey and answered the questions as honestly as he could, but some of his responses seemed far-off the mark. For instance, when asked about special talents he replied: *none, except for business, as evinced by keeping accounts, replies to correspondence, and investing money very well. Very methodical in all my habits*. So, he blithely ignored all his contributions to natural science he had made and his book on the *Origin of Species*. However, he was spot on the target when asked whether he had any strongly marked mental peculiarities: *Steadiness, great curiosity about facts and their meaning. Some love of the new and marvelous*. Galton tactfully declined to ask him about the nature of heredity and how it worked. (See Epilogue 3 for some more views.)

Years later Galton still talked about Darwin's theory of inheritance as though it was true. Galton gave a lecture to the Anthropological Institute, subsequently published in the *Contemporary Review*. His talk was titled "A Theory of Heredity" and in it he virtually gave a word-for-word account of Darwin's theory with some minor modifications of his own. He started the lecture along these lines:

> *I have taken the subject of Natural Inheritance on this occasion and were it left to my own will, I would prefer to repeat it every year—so abundant is the interest that attaches itself to the subject, so wonderful are the varieties of outlets which it offers into the various departments of natural philosophy. There are no phenomena in the biological world that are not touched by the laws of natural inheritance. There is no better, there is no more open door, by which you can enter into the study of natural philosophy than by considering the phenomenon of heredity in all its aspects. I trust I therefore will not disappoint you in choosing this for my subject rather than a newer topic which could not be better, were it even so good. Let us now try to trace distinctly the connection between certain bodily features of an adult and how they are passed on to the offspring.*

He then went on to state these three propositions: (1) that hereditary units occur in great numbers; (2) the germs or gemmules of such units occur in still greater number and variety (existing somewhere); and (3) undeveloped gemmules do not perish but multiply and are transmissible. He eventually proposed his *laws of ancestral heredity* that require the contribution from more ancestors than the parents to the inherited features of the child, and definitely implied the blending of inherited characters.

Naturally, Darwin was delighted when he started to read a copy of the text of the lecture; *Good for Galton; he is still sticking up for the gemmule theory*. As he read on he became more and more confused as undefined terms were introduced: "germs" were used instead of gemmules; Galton introduced words such as "stirps," "developed germ," "residues," and so on. Galton was not sticking to the Down House interpretation. So much so that Darwin admonished his cousin by a letter:

> *I have read your essay with much curiosity and interest, but you probably have no idea how excessively difficult it is to understand. I cannot fully grasp what are the points on which we differ—I dare say this is chiefly due to muddle-headedness on my part, but I do not think wholly so… unless you can make several parts clearer, I believe (although I hope I am altogether wrong) that very few will endeavor or succeed in fathoming your meaning….*[24]

Probably Darwin thought this lecture was just playing around with a new jargon on which one can impose any definition that one wants.

Even more extraordinary was the second edition of Galton's book on *Hereditary Genius*. It came out in 1892, that is, 10 years after Darwin's death, and he left Darwin's theory of inheritance in the last chapter virtually intact and ignored his own ideas on the subject. Thus in his own words:

> *This theory [of gemmules] propounded by Mr. Darwin as provisional is … of enormous service to those who inquire into heredity. It gives a key that unlocks every one of the hitherto unopened barriers to our comprehension of its nature…It is very advisable that we should look at the facts of hereditary genius from the point of view which the theory of gemmule affords…*

In the preface to this edition of the book he did mention that Darwin's theory should require revision; however, he seemed unable to bring himself to do it. He believed it to be a marvelous theory and having remarkable powers to bring large classes of apparently different phenomena under a single law.

So much do our personal relationships interfere with our ability to appreciate the truth. It seems he could never shake off the dominance that Darwin had exerted over him for so many years and it persisted long after Darwin's death. To quote yet again Galton's lecture to the Royal Society in 1886:

> *I rarely approached his [Darwin's] general presence without an almost overwhelming sense of devotion and reverence and I valued his encouragement and approbation more perhaps than the whole world besides. This is the simple outline of my scientific history.*

It was even stranger that Galton, having been led onto the wrong path by Darwin, decided to follow up Darwin's later advice in 1870s to explore the laws of heredity in sweet peas by working out what quantitative features of the parent seeds, such as diameter and weight, were transmitted to the next generation of seeds. In many ways Darwin's advice was good, as the sweet pea is easier to work with than rabbits. Darwin might have recalled the reprinted article that had been sent from Brünn. The monk had been working on the edible pea but the sweet pea might be better. They are hardier, they do not have a tendency to cross-fertilize, so that the seeds can grow near differently colored plants in neighboring beds; and the size and number of seeds per pod are more constant; this would give rise to less variation in the measurements than for edible peas. The pod of the edible pea contains many more seeds that get smaller toward each end. Measurements would be much easier with the sweet pea. Galton followed up this line of enquiry and started a set of experiments that bear an uncanny resemblance to Mendel's experiments with the edible pea. He obtained a large number of sweet pea seeds and sorted them into seven groups, K being the largest in size (regarding diameters and weights) to L, M, N, O, P, and Q being the smallest. He planted each group in a bed that was 1.5 feet wide and 5 feet long with 10 holes for each of the 10 seeds. The beds were bushed to keep off the birds. As the seeds of the next generation became ripe, they were collected, measured, and compared to those of the parental generation. He then calculated what he called the index of correlation being a fraction between 0 and 1 based on the agreements of parent to offspring measurements that he called r; this connects the parent pea to the offspring pea measurement. He found that when there is no relationship at all, r becomes zero; when the measurements are identical between generations, r becomes 1. Alas, he was still thinking along Darwin's ideas of "more or less" variation and not of Mendel's idea of "all or none" variation that stands out clearly such as tall or dwarf plants. He was not looking for a numerical basis of inheritance but trying to see if there were any statistical correlations between the features of seed weight or size between the parents and offspring that one might expect if blending inheritance were at work. His correlation index is still used today after much modification as the correlation coefficient, to measure such similarity between variables. He plotted the mean diameters of parent seeds against the mean diameters of the seeds from their offspring and reached the idea that the slope of the line would measure the magnitude of the inheritance between the parents and offspring. If there was no slope, the diameter of the offspring pea would be unrelated to any diameter of the parent pea; if the slope were to be 45 degrees (i.e., a slope of unity), the diameter of the offspring peas would be equivalent to that of the parent peas, therefore indicating a one-to-one correspondence. He went on to use this idea of correlation of the mean of midparents height to mean sibling heights to study the inheritance of human stature and found the slope to be 0.33. He also used a similar approach to assess the inheritance of eye color in 4490 individuals from 168 three-generation families with not very convincing results.

Although he was doing much the same experiments as Mendel, the ideas driving him at the start of the enquiry would never lead him to replicate the results of Mendel's experiments. If only he had chosen to study the shapes of the pollen grain that in the sweet pea are discrete, either round or oblong; this being a qualitative trait quite unlike Darwin's chosen study of quantitative, or continuous variation. The pollen shape is inherited similar to a Mendelian trait, so very probably he would have found 3:1 ratios similar to what Mendel had done 5 years previously, and

Bateson was to find using the sweet pea pollen grains about 20 years later. He no doubt felt like a fool when Mendel's ratios were rediscovered at the beginning of the twentieth century at how close he had been.

However, he had developed a very interesting and original statistical work on correlation analysis with his model before Darwin died. His correlation test to define relationships between two variables, after much refinement by statisticians such as Pearson, Spearman, Kendall, and others, has been in use to this day. Galton would have done better to get away from Darwin's influence from the start of the study and to use some of the newer ideas he was developing of "stirps" and "germs."

Romanes also followed up Darwin's suggestions on gemmules but in a different way and under Darwin's personal supervision.

Meanwhile, what path was Mendel pursuing? Abbot Cyril Napp aged 75 had died in 1868. The monks now had to appoint a new abbot. In a close election with his fellow priest Tomas Bratanek, Mendel was chosen to the prelacy. *From the very modest position of teacher I find myself moved into a sphere in which much appears strange to me*, he wrote to von Nageli, *and it will take some time and effort before I feel at home in it*. This high ecclesiastical office may have been more important to Mendel than the position of a naturalist and a scientist. His work was now to lead the monks as their religious superior, not only in the spiritual sense but also as their judge under Canon law. For legal disputes he would be in most part answerable directly to the Pope, being of equal standing to the local bishop. He would be treated with the utmost reverence by the brethren of the monastery; when he appeared in a church or chapter wearing the pectoral cross, all present would rise and bow. No monk might sit in his presence or leave from there without his permission. The highest place was assigned to him both in the church and at table. He apparently filled all these duties and obligations with due humility judging from his obituary of 1884. All this ceremony would be anathema to most scientists who would think Mendel to be demoted to an administrator or manager. It is true he had many civic duties to perform as an abbot. He would automatically be involved with committees of the local school authorities and scientific societies that took up much time.

Then as a prominent citizen there were numerous social occasions to attend. Local dignitaries would come to the monastery to discuss civic matters. Councilors Januschka and Ruber; Dr. Scharrer, president of the Supreme Court of Moravia and councilors Schilda and Strobach; and Professor Rost of the Brünn school were all recorded as visitors.

Mendel drops out of the heredity story from now until the start of the twentieth century when his work was rediscovered.

chapter eight

Finding allies

> *Disinterested intellectual curiosity is the life-blood of real civilization.*
>
> **Areopagitica. J. Milton (1608–1674)**

Galton contacted Romanes to ask whether he would still like to continue their collaboration with further experimental work on the subject of heredity. Galton had developed some new and provocative ideas and wanted to discuss them with Romanes who sometimes invited him to stay for a long weekend at his rented country house, the Geanies, in Ross-shire, Scotland. It was a beautiful old house overlooking the Moray Firth. The countryside around, without being romantically beautiful, had a charm of its own. There was a certain melancholy and loneliness about the landscape that appealed to Romanes in October days of perfect beauty, which seemed especially peculiar to Scotland. It abounded with every type of seabird and it was almost impossible to describe the strange twilight of the summers, the silence only broken by the hooting of owls and the scream of the seagulls. It was an ideal setting for a poet, a naturalist, a botanist, and a sportsman, all of which in modesty Romanes considered himself to be.

Romanes's happiest days had been spent at Geanies tramping over the moors and plodding across lots of turnip fields, a major crop for those parts. Romanes often took his guests out for hunting on Saturdays with their gamekeeper. They tramped for hours together across more turnip fields and grassy meadows. They often bagged up to twenty brace of pheasant, a brace or two of plover, and other game such as hare or duck. They could have easily got more, only Bango their setter would get so tired by the afternoon that they usually went home at five-o'clock. Their new setter Flora was a beauty. She was very much like a Bango but with a prettier face and she was a splendid worker. They marched back home across more turnip fields—a weary, happy party. Turnip fields always reminded Romanes of a letter that he had written to appease his social conscience perhaps of inheriting so much wealth from his father. It was to the editor of the *Times* and read:

> *Dear Sir,*
> *A weakly looking local lad, aged 17 was charged with stealing two turnips; value 3 pence, growing in a*

turnip field belonging to Mr. H. Bunce. The plaintiff having lost a quantity of turnips previously had set Police Constable Whitty to watch the property and he saw the prisoner pull the two turnips and put them in his pocket. The accused said he had had nothing to eat all day, and being very hungry he took the turnips. A previous conviction was proved against him for felony and he was now committed by Mr. Denham, the judge, to six weeks hard labour.

One would like to possess a good large field of turnips where each turnip can be fairly valued at 1 ½ pence. But, taking this as the true value of the particular turnips in question, it appears that a starving man is now serving a week's hard labour for every half penny's worth of the cheapest possible kind of food that he could steal. It is, of course, very right that he should have received some measure of punishment, if only to be a warning to others in the neighborhood; but the measure of punishment, which he did receive, seems, in the face of the matter monstrous. We are not told what was the "felony" for which the weakly looking lad was previously convicted; but at any rate we do know that on the present occasion his theft was not for any purpose of pecuniary gain. It must have been, as we said, merely to alleviate the pains of hunger, or otherwise he would have carried some more capacious receptacle than either his pocket or his empty stomach. On the whole, therefore, I say—and say emphatically—this case demands some explanation.

The letter was published and Romanes, somewhat indecisively, never followed up the case to see whether it had the desired effects.

Geanies was a large rambling house with long passages and mysterious staircases. The library was a most lovely room lined with bookcases and leading into an old-fashioned garden full of flowers planted in long herbaceous borders. While at Geanies, Romanes always liked to do some serious work in this room. He might prepare an academic paper, or review a paper for *Nature*, or correct the page proofs of a chapter he had written, or write an essay on such subjects as *Freedom of the Will*. One time while correcting proofs of a book, his little niece came in wanting to play the *gee-gee*. Romanes said, *No dear, Uncle is writing*. She asked, *writing letters or writing book?* He said *writing a book*. Whereon she made the shrewd remark—*Uncle not writing to anybody, uncle can play gee-gee*. So much for her estimation of his popularity as an author.

If tired he went to his study where he copied Darwin's tastes for reading simple pure romances such as the novels *The Heir of Redclyffe* and *Chaplet of Pearls*. They often brought tears to his eyes; and he had read the last one many times over.

The present meeting, however, was to be given to Francis.[25] He started to explain that the gemmule hypothesis, because of both their studies, was now most likely to be wrong—but perhaps a modification of it may turn out to be correct. His exciting new idea was that the hereditary material might be collected and organized into *stirps* (from the Latin *stirpes*, a root) and each stirp contained the sum total of all the great variety of germs (or gemmules as Darwin would call them). You could think of stirps rather like corn-on-the-cob; in which the corn seeds are the hereditary material and organized in a particular way on the cob. However, the cobs only reside in the reproductive organs (ovary and testis) and are not found anywhere else in the body. The cob seeds direct the development of all the body cells and determine the final appearance of each adult organism. Since they are only lodged in the reproductive organs, it is very unlikely that bodily alterations of an individual due to environmental factors could be transmitted to the offspring, because they would first have to alter the composition of the cobs in the sex organ. Therefore, inheritance of acquired characters would be a very rare occurrence and would have to alter individual seeds on the cob tucked away in the ovaries and testicles. This would be directly against Darwin's view that allowed for variation in the bodily organs such as the muscle or skin to affect the number of gemmules they shed for traveling to the sex organs for onward transmission to affect the inheritance of the next generation.

Galton's hypothesis of cobs and corn seeds would allow him to specify with much greater clearness the curious connection between the characteristics of the offspring with the parents. The idea of it being a simple one-to-one descent is wholly untenable, and is the chief reason why most people seem perplexed at the appearance of variability in hereditary transmission to offspring. The cob of the child should be considered to have descended directly from two cobs: one from the father and the other from the mother; and it is this mixture of cobs that gives rise to the individual variability of the child, representing some features of each of the parents. If a reduction in cobs from two in the parent to one for the child did not occur, the children of later generations would go on accumulating more and more cobs *ad infinitum*, so there must be a reduction or suppression at some stage. The seeds in each parental cob may compete for survival because the cob from which the child springs must be only half the size of the combined cobs that he receives from the two parents. Thus one half of a child's parentage must be in some way suppressed. How could this happen? Galton did not know but compared the cob to a nation of individuals where only the foremost men of that nation become its representatives; so likewise with the seeds on the cob, only the fore-most seeds succeed in representing each of the parents in the offspring. In animals of pure breed, whose cobs contain one or only a few types of seeds, the offspring will always resemble their parents and each other; the more mongrel the breed the greater will be the variety of the seeds on the cob and hence the variety in the offspring. This may appear a fanciful metaphor, and he was not trying to be particularly poetical but only try-ing to make his ideas more vivid. These metaphors can be absolutely nec-essary for the study and communication of new ideas; they can portray a common element (or relationship), which the two different things may possess. It allows one to anticipate what one might expect to find. In trying to visualize the unknown, the imagination must clothe it with attributes analogous to what we already know, and this gives rise to the productive metaphor. Imagination conceives of ideas that hitherto had passed unno-ticed. Before we knew the nature of electricity it was convenient to think of it as a fluid. This metaphor suggested many fruitful analogies such as dif-ferences in level in a battery, direction or reversal of flow, resistance to flow, leakage to earth, and so on; all of which nearly corresponds to the move-ment of electrons once they were later discovered. Any period of rapid and extensive development in a scientific field has to be associated with episodes of loose thinking as one grapples to formulate new concepts and explore all the possibilities. The perception of a new truth (in genetics) can often start from the conception of such a metaphor or analogy as seeds on a corncob. This idea of seeds stacked on a cob is very close to our present day concept of genes lining up on nucleosomes (like beads) on a chromosome.

Although Romanes thought Galton's ideas sounded exciting, was he really taken with them? Galton's ideas involved a complete rejection of Darwin's gemmule hypothesis and could be construed as some sort of disloyalty against Darwin's authority. In view of Mr. Darwin's high repu-tation and his kindness to Romanes in the past, he was rather loath to take any part in this. Galton's ideas would imply a continuity of the hereditary material from parents to offsprings for many generations, which inciden-tally was very close to the theories of Professor Weismann (1834–1914) that Romanes had heard about. The professor postulated the stability of the "germ plasm" (that is the hereditary material, which we now know to be DNA) from generation-to-generation. Incidentally, Professor Weismann acknowledged this in a letter to Galton writing:

> *It was Mr. Herdman of Liverpool who—some years ago—directed my attention to this paper of yours … I regret not to have known it before, as you have exposed in your paper an idea which is in one essential point nearly allied to the main idea contained in my theory of the continuity of the germ plasm.*

Again this idea went directly against Darwin's views. It also went against Darwin's opinions on the inheritance of acquired characters. It would mean that nothing that happens in the lifetime of the individual could exercise any influence on the germ plasm in the sex cells or on the subsequent progeny. Effects of use or disuse of a body part cannot be inherited; nor can any other adaptation to environmental conditions be transmitted to the offspring by altering the germ plasm (DNA). In Galton's view, natural selection could only operate through spontaneous variation occurring in the seeds of the cob located in the sex cells, and only those variations were used that in the resulting adult were best suited for its survival and reproduction, so therefore transmitting to the next generation. These again were not Darwin's opinions. So, Romanes asked Galton to give him time to think over the matter and to consider some of the experiments that they might perform together to test these ideas.

On the Sunday morning, as was Romanes's usual practice if no clergyman was of their party, he conducted a short service for all their guests and for the household servants who could not get the ten miles to the nearest church. Although Romanes fully accepted the doctrine of evolution, he still held strongly to his Christian faith. Underneath his obvious love for scientific work, there was always the same longing and craving for the old religious beliefs. He used the question: "Is Christian faith possible or intellectually justifiable in the face of scientific discovery?" for one of his popular sermons. In London, Romanes regularly attended church, usually Christ Church in Albany Street where the future bishop of St. Albans was then vicar. At the end of his service at Geanies, Romanes always gave a sermon and he usually took it from his published book on a *Candid Examination of Theism.* Galton as a guest would have to attend, no doubt being rather skeptical throughout the proceedings—perhaps because he thought Romanes preached for too long or that the contents were nonsense.

After lunch Romanes invited other neighbors across to the house and they all would have a very merry time, with party games, amateur theatricals, and heated discussions about nothing in particular. One game that caused much amusement was who could best "card wool" in opposite directions; that means turning the right hand round and round one way, whereas at the same time turning the left hand round and round the other way. This was always popular and Romanes enjoyed seeing his guests winding their hands into knots and reach a climax when a neighbor often ended by spilling his glass of wine off the table into his lap. Galton entered freely into the spirit of the Geanies brotherhood and told some excellent jokes that made them all laugh heartily. He mentioned that when he was told that Miss Barrett had married Mr. Browning he replied: *It's a good thing these two understand each other, for no one else understands them.* Romanes prided himself on his own jokes, but unlike Galton's they were free from unkindness and he did not use repartee or epigram, the point of which often lay he thought in malice.

Here is one of his best stories. *A Rector was asked to take the chair at a prayer meeting. One of the parishioners prayed as follows: O Lord, we had a sermon from our vicar yesterday and we thank Thee for it because it was an able discourse, but we pray Thee to give him some idea of what the Gospel is!.* He thought this joke killingly amusing don't you think.

Their party broke up in high spirits on Monday morning when their guests departed.

chapter nine

Still chasing the truth

... to save you from being talked into error by specious arguments.

Thessalonians 2:3

The long greenhouse at Down House built alongside the south-facing wall at the rear of the estate was a favorite place for Darwin to take his visitors to discuss the burning issues of plant biology.[26] The central part of the building was heated to simulate tropical conditions. Darwin had done experiments to find out if seeds were viable after prolonged soaking in seawater; and could they survive to colonize islands such as the Galapagos after drifting to them in the sea?

In the cooler part of the building were trays of potatoes stored under the bench. He had been studying the properties of grafted potatoes to see if they supported his gemmule hypothesis. Darwin and Romanes discussed Galton's recent ideas on inheritance. Romanes had read several of Galton's papers on a new theory of heredity. He said, rather two-facedly, that they appeared to him to be quite destitute of intelligible meaning. It was a jumble of confused ideas on heredity couched in a hotchpotch of new jargon, very ill defined. Darwin tended to agree—he had some more correspondence with Galton and the confusion was even more confounded with respect to the points in which they differed. Darwin evidently disliked the new ideas of his young cousin. So, how should they proceed with the gemmule hypothesis?

Darwin suggested that Romanes should continue working with him. If it had to be a straight choice between Galton and Darwin there would be no doubt who Romanes would choose. Darwin had virtually made Romanes' career up till now.

Darwin suggested a few possible lines of attack. Now we all know that the world is more convinced about the truth of any matter by experiments on animals. However, there is still a place here for work on plants. A large number of successful results in any field will help to convince people. Darwin's new idea was one he got from a very remarkable case report given in the *Gardeners Chronicle* of January 2, 1873. A vine was grafted onto a different variety of vine and the host plant took on some of the features of the grafted plant. This clearly supports the idea of gemmules passing across from the graft and influencing the development of the host plant.

95

Now why could not they try a series of systematic experiments using grafts of one species of plant, such as the potato, onto another? Darwin thought that the potato would be better experimental material than vines because pieces of tuber can be inserted quite easily into the host potato tuber. With Romanes' energy and skill in experimentation he would be sure to be successful. It would be absolutely splendid if Romanes could see if it would work. Darwin also mentioned that Galton was much less skeptical about the gemmule hypothesis than he was immediately after those disastrous rabbit experiments—so, Darwin thought that they might be able to win him round.

As it happened Darwin wanted to put Romanes up for membership of the Linnaean Society[27] at the same time that he was proposing to do the same for his son Francis. With Darwin's support there was no doubt that Romanes would get elected. Romanes had a Cambridge MA degree and was a fellow of the Philosophical Society of Cambridge. His published papers were a bit thin on the ground but Darwin did not think that would count against him. With such encouragement from Darwin, Romanes decided to stick as close to Darwin as possible, to take up the gemmule problem again with gusto, and to follow his suggestions to the letter.

Romanes valued Darwin's opinion in everything; he always found his judgments more deep and sound than all others.

The initial experiments were to try grafting the crown of the tuber of a red-skinned variety of potato into the eye of a white- skinned potato, and vice versa; then to examine the progeny to see if they have taken the color of the grafted variety.

Romanes began with a method that he thought very cunning. It was to punch out the eye of the donor potato with an electroplated cork borer and then place it in a flat-bottomed hole of a slightly smaller size made with another cork borer in the host tuber. The fit was almost perfect. Luckily the inserted plugs adhered quite well and he got about 100 potatoes planted out. A great many potatoes came up and he was excited to see what would happen. He resolved not to leave the gemmule hypothesis alone from now on until he was quite sure that it could not be validated by any other type of experimental work.

Next year Romanes went to visit Darwin to give him some of the results. He had dug up all the potatoes and some of the produce looked suspicious, although more than this he would not dare to say. The batches marked A and B were the controls being the original varieties of potatoes before any grafting. The rest of the batches were the results of grafting. Batch C was the oddest and to Romanes perhaps too partial, eyes looked very much like a mixture of characters of the two varieties of potato. In the case of batch D many of the potatoes were rotten, so it was difficult to tell. Romanes wished that he had begun these experiments a year earlier to have the results ready for the second edition of Darwin's book on

The Variation of Animals and Plants under Domestication. Darwin was already correcting pages for this second edition and he would certainly like to have included a short abstract of Romanes' work on potato graft hybrids. It was to be published next November. Darwin had collected all the other important examples including the one about the vine grafts in *Gardener's Chronicle*. Darwin with due caution had thought the results in the vine could just be due to bud variation and not have anything to do with the effects of the grafting. He asked Romanes to give him a concise summary of the conclusions from his work so far.

Therefore, Romanes wrote this summary for him:

> *The experiments in graft-hybridization prove that formative material (or gemmules) are actually present in the general tissues of the plants and are capable of uniting with the gemmules of another plant and thus reproducing the entire organism. The hybrid appears to present equally the characters of the graft and host showing that the formative material (gemmules) must have been present in the tissues of the graft and had an effect on the development of the host. These facts are fully in harmony with the theory of gemmules.*

Romanes was so enthused by his own conclusions that it made him more anxious than ever to get further positive results using grafts for beans, onions, carrots, dahlias, and peonies. There seemed no doubt to him that such results must be obtainable if one hammered away long enough at the problem.

Darwin wrote to Romanes later in 1876 wishing him all success in further work testing his hypothesis. He told Romanes that Trollope, in one of his novels, gives as a maxim of perseverance by a brickmaker that: *It is dogged as does it*. And he told him that this should be the motto for his own projects on grafting. Darwin adopted here the somewhat dismal fallacy that perseverance is the measure of achievement. However, it is always very difficult to know when a research project should be dropped; the next set of results might turn out to be critical for verification of the idea.

As Darwin kindly proposed Romanes for election to the Linnaean Society, Romanes had the opportunity of doing a similar favor. In 1876 Romanes was the secretary of the Physiological Society and he got Darwin elected as the first honorary member. It was not difficult. The then president, Dr. Michael Foster said: *Let us pile on him all the honors we possibly can*. It seemed to Romanes that Darwin never fully realized the height of his pedestal, so that he was glad of any opportunity of this kind to show Darwin at the angle which all their upturned faces should be inclined, that is looking nearly vertically upward.

Two years later, Romanes gave a lecture on the topic of heredity at Glasgow University. The hosts first gave a dinner in Romanes' honor; the guests assembled being the most important men in Glasgow. Sir William Thomson, later to become Lord Kelvin having formulated some of the laws of thermodynamics, was the biggest name there and Romanes spent most of the dinner talking with him. *The advantage of meeting celebrated men when oneself is also a celebrated man* (how sweet is self-contentment) *is that the two know all about each other before they meet and are friends from the start,* as Romanes later wrote in his memoirs. They then went to the lecture where Sir William took the chair and introduced Romanes to the audience with such a glowing oration that it almost took him by surprise. The audience was thus led to suppose that Romanes was one of the brightest of all bright scientific stars in the firmament and so welcomed him very warmly. Romanes got so enthusiastic that he discarded his notes and lectured freely in the most magnificent style, even for him. This is the highest praise he could bestow on himself. He spoke for an hour and a half and told a number of jokes that did not appear in the printed lecture. He never heard an audience laugh so much. For good measure and as a grand finale he brought in Darwin's name as his collaborator in these studies. He expected an outburst of applause, but the loud and long-continued cheering beat anything that he had ever heard before. At the end, he bowed to the audience twice and would have done more bows but for fear of making himself look ridiculous. Afterward many eminent men in the audience had so much praise for him, and Professor Caird went so far as to say that it was the most successful lecture he had ever heard. It was really enough to make one quite conceited.

The vote of thanks was proposed by Professor McKendrick and Romanes was met by another storm of applause, so that he began to feel quite overcome. He managed a few words with all becoming modesty, and then Sir William summed up the proceedings. Romanes gave the same lecture several times again (in Leeds, Birmingham, and Dublin) and they always went off extremely well in his own opinion.

Some years later (in 1879) Romanes was married on February 11 to Ethel, the only daughter of Andrew Duncan Esq. of Liverpool. He met her at the house of her cousin and guardian, Sir James Malcolm of Balbedie and Grange, Fifeshire. It was a brilliant match, even if Romanes himself said so. He was at the peak of his career and she was a lovely, intelligent, and wealthy young woman from a good family. She gave him many years of a bright and most happy domestic life.

The married couple used to give grand dinner parties, in all the glory of their new mahogany furniture and silverware. After dinner they would adjourn to the drawing room and two of their friends would play enjoyable duets at the piano. It was at about this time that he began to take up poetry seriously, inspired by the love his wife gave him. He managed to publish some of them in lesser-known magazines. He of course wrote under a pseudonym, as he naturally did not want to mix his poetry with his scientific work. People might become less serious and convinced about his academic publications if they thought there was an element of fancy or poetry in them. He did not want it to be known that he had this additional talent and he insisted that if his verses were to appear in any publication related to him in the future then he would require them to be without his name and perhaps he could modify any of the lines that might lead to the author being identified. An example of his verse to commemorate Darwin's death is given in Chapter 11.

His wife thought that some of the poems were so good that he had them personally printed for private circulation. He had one collection bound as a grand presentation copy to give to his dearest wife as a memorial for their tenth-year wedding anniversary; it included a special sonnet dedicated to her. He also sent copies of his book to Mr. F. Palgrave of "Golden Treasury" fame, to Lord Tennyson, Mr. Edmund Gosse, Mr. George Meredith, and to Mr. W. E. Gladstone. He received very kind comments from them particularly on his odes and his poem titled the "Dream of Poetry." The Rt. Hon. W. E. Gladstone sent him a long congratulatory letter finding it a most acceptable and considerate gift. He wrote that Romanes obviously had a very considerable poetic gift and that he could see no reason why a man of science should not be a good poet too. He quoted Lord Bacon (1561–1626), a scientific philosopher and essayist, pointing out that his essays had much of the poet in them. With this encouragement, Romanes grew more and more addicted to versify toward the later years of his life; and ladies who became more intimately acquainted with him were sure to have, sooner or later, a sonnet sent to them on some special occasion.

In the same year that he married another good fortune befell him. He was elected to the highest scientific honor of the land as a Fellow of the Royal Society. No doubt this was partly due to his work on gemmules following the theories of Darwin, who was already a Fellow.

chapter ten

Losing allies

Be strong and of a good courage, fear not, nor be afraid of them.

Deuteronomy 31:6

A common place for Romanes to meet Alfred Wallace, the codiscoverer of natural selection and a great supporter of Darwin's views on heredity, was at meetings of the Linnaean Society. The Linnaean Society was founded in 1788 for anyone interested in biological matters and contained a large part of the botanical and zoological collections of the great Swedish naturalist Carl Linnaeus (1707–1778) who was the first to provide a rational and systematic way of classifying the members of the plant and animal kingdoms. This was the learned society where Wallace had presented his theory of natural selection conjointly with Charles Darwin in 1858, the papers being read to the audience by Sir Charles Lyell and J. D. Hooker. The entrance to the society is through the gateway to Burlington House in Piccadilly, London and the building also accommodates the Royal Academy for Arts and Antiquaries. The Linnaean Society has a poky little entrance hall, with a glass case containing various memorabilia and small cameo portraits of Linnaeus. A spiral staircase hung with portraits of various celebrated botanists leads up to the main meeting room on the first floor.

Romanes and Wallace were on quite frosty terms (due to arguments about spiritualism, described in Epilogue 2). Romanes knew that Wallace had in the past been a fervent supporter of Darwin's gemmule theory but feeling in the need for some encouragement (all his grafting work had been negative) he started telling Wallace about some mutilation experiments that he had been doing. Romanes had heard of no less than three cases of cats whose tails had been cut off earlier giving birth to kittens that were also tailless, presumably having run out of tail gemmules for the kittens. He had asked his assistant to procure either Angora or Persian cats and was trying to repeat these experiments.

Wallace considered that he was wasting his time, it would never work. A German professor, August Weismann (1834–1914) of germ plasm fame working in Freiburg, had already tried it. With German thoroughness he had cut off the tails of white mice repeatedly over five generations of their progeny. 901 young were produced by five generations of artificially mutilated parents, and yet there was not a single example of a rudimentary tail or of any other abnormality in this organ. They all grew normal tails. Anyway Wallace had gone off the idea of gemmules. He initially accepted the hypothesis piecemeal because it could explain the inheritance of acquired characters, such as the effects of use and disuse on body parts. He felt relieved at having at least some hypothesis, however, provisional, that would serve to explain the facts. He told Darwin *I shall never be able to give it up till a better one supplies its place.*

He had now found a better one.[28] Wallace grew convinced by Galton's experiments on rabbits that opposed the gemmule idea. In every case the offspring resembled the biological parents and never the rabbits that supplied the blood for transfusion. There were other critics who had pointed out that it is very well known that if you take a plant stock onto which a different variety is grafted, it ought to according to the gemmule hypothesis change the character of the fruit produced by the host to resemble that of the graft—but it never does.

Also this idea of Lamarck's (1744–1829), of "use or disuse inheritance," that you can inherit the bodily features from your parents that have been produced in them by environmental factors; Wallace thought that this was quite wrong too. Take the case of a strong muscular father working as a blacksmith or carpenter, Lamarck believed that the sons would very likely inherit the same degree of muscularity too and turn out good blacksmiths or carpenters. Darwin obviously believed this sort of thing and framed his gemmule hypothesis to account for it. The strength acquired by the arms of the father by constant exercise would be transferred to his son by the excess release of gemmules flowing from the father's arms to his reproductive organs and then transmitted on to the son where they could develop a strong musculature for the child. Darwin used this to explain the effects of use or disuse of parental features on the future progeny. There was absolutely no trustworthy evidence that this occurred; and there was plenty of evidence against it. The anecdotal evidence for the inheritance of acquired characters did not carry much weight with Wallace. He scoffed at the idea, for example, that the origin of new features such as the horns on deer and cattle arose from the habit of continual butting with their heads leading to thickening of the skin, callous formation, and finally to excrescences of bone to anchor the horns firmly on the head. The horn size then increased by persistent use to produce all the great variety of horns one can find in the ungulates. In fact, Wallace believed it happened in reverse, that chance variation had given rise to bony bosses on the foreheads of hoofed animals and these had later evolved into horns by sexual selection. Deer or antelopes with the strongest horns would give rise to more offsprings with the same feature. Indeed, careful study had shown that projections on the frontal bones of ungulates could be found as an occasional variation in some species that never develop horns, such as the horse. No known animal in the ancestral line of horses ever had horns, so this must have been a newly acquired character by chance but never evolved into horns because horses have such powerful weapons of offence in the form of hooves.

In Wallace's view, Francis Galton had suggested a much better theory of heredity. It was a *decided improvement over Darwin's as it gets over some of the great difficulties of the cumbrousness of his ideas.* Moreover, it had the backing of some experimental evidence. Galton's corn-on-the-cob hypothesis was something really original and, within its very limited range, as important as possibly leading to new concepts regarding the laws of heredity. It was only misleading when trying to account for the *Origin of Species*, which was always, in Wallace's view, based on natural selection alone. Galton appeared to have no adequate conception of Wallace's views on what natural selection was, how it worked, or how impossible it was to escape from it.

Galton had stated that "discontinuous variants" or "sports" could possibly provide the basis for evolution by providing a survival advantage to some members of the species who possessed these chance variants. According to Wallace this was quite wrong. On account of the extreme constancy and severity of elimination of individuals through survival of the fittest (a famous term originated by Herbert Spencer), such abrupt variants could never become permanently established in a breeding population and so could play no role in its evolution. Incidentally, Wallace also took exception to Galton's new ideas about eugenics, that the human race could be directly improved by artificial selection for breeding of superior individuals. His main objection being that natural selection was so constant, universal, and so powerful to include a spiritual component that no forms of artificial selection could ever overcome its effects.

The "germs," or as Wallace now liked to call the hereditary particles in the sex organs of each individual, do not come from any of the bodily structures during growth and development of the parents, but pass directly from the parent's own hereditary particles to the offspring. They are not produced anew from the various body parts, but reside as a store or bank of particles in the reproductive organs of the parents that in turn came from their parents and then their parents before them and so on. Wallace had adopted the similar hypothesis proposed by the German professor, August Weismann which he called his theory of the continuity of the germ plasm. Weismann got the idea from his embryological researches that the sex cells of animals contain *something essential for the species, something that must be carefully preserved and passed from one generation to another for the embryo to develop.* This material was contained in the cell nucleus. It was all rather speculative at first but he made several important predictions that were found to be correct. As the hereditary substance from two parents becomes mixed together in the fertilized egg, there would be a progressive increase in the amount of hereditary material unless at some stage there was a compensatory reduction of the nuclear material. He therefore predicted that there must be a form of nuclear division in the parents' sex cells to halve the amount of ancestral hereditary material to be passed on to the offspring. The cytological work of other investigators demonstrated the correctness of this prediction and found that the chromosomal numbers were indeed halved in the sperms and eggs. This gave rise to the idea that the hereditary material could be carried on the chromosomes, structures first observed and studied in 1881 by E. G. Balbiani. Such predictions turned out to be correct, and were confirmed by much other evidence. The idea immediately came to be accepted by most biologists in every part of the world; and it came very close to Galton's ideas on stirps. However, it took until the 1940s before it could be unequivocally shown that this nuclear material was DNA and carried the instructions for a cell to make proteins; the reduction of the amount of DNA at cell division was observed earlier in the late 1880s during the process of meiotic cell division in sea urchin eggs.

What it implied was that changes produced in an individual during life by exercise or use of his senses such as vision or any other environmental agents cannot affect the inherited material transmitted to the offspring. A giraffe's long neck does not arise because the animal stretches its neck to reach the top leaves of a tall tree and then passes on this adaptation to its offspring; but because a chance variation in the ancestral giraffe produced a longer neck that allowed it to feed more efficiently on tall trees, and so produced more offspring than other ancestors with shorter necks confined to browsing from smaller trees. What is inherited is the capacity to develop into a form more or less closely resembling that of the parents (or their direct ancestors), and the same features appear in their offspring, uninfluenced by the environmental conditions; thus leading to all that wonderful variety of species that we see in the biological collections of Carl Linnaeus.

Romanes' heart must have sunk. He left such meetings feeling discouraged, another disciple lost to the cause; and he fervently hoped that all the other biologists of Europe were not being persuaded over to Wallace's camp. Surely Darwin's great reputation still counted for something. He resolved to go on earnestly seeking for more facts that would serve as a crucial discriminatory experiment for the rival theories: gemmules coming from all the cells of the body, or discrete particles (genes) for inheritance only coming from the sex cells.

chapter eleven

Darwin's and Mendel's death[29]

I am old and I do not know when I may die.

Genesis 27:2

A few years after Romanes had performed more gemmule experiments Darwin came to visit him at Cornwall Terrace, Regent's Park in 1881. Darwin was staying in London for a week with his daughter in Bryanston Street. He had written to Romanes the week before about other experiments that should be done on his gemmule theory and wanted to explain the best approach for this. Romanes wanted to see him about a different matter regarding Francis Galton. He had been wondering why Oxford or Cambridge had not offered Galton an honorary degree in view of his excellent work and was curious to know whether Darwin would start a movement in that direction.

Although Romanes had spent more time and trouble than he liked to acknowledge in trying to get evidence to support Darwin's theory, he never obtained any positive results. Following Darwin's advice he did not care to publish negative results, so there are no papers of his on the subject, although he fairly believed that no other person had tried so many experiments. Apart from the tailless cat experiments, he had spent one year mutilating caterpillars at the zoo to see if such deformities could be inherited. The caterpillar is virtually at the embryonic stage of the development of the butterfly and the defects introduced at this stage might be more readily transmitted. He was not cruel and only removed a tuft of red hair at the front end of each of the larva; he never found any evidence that the deformity was transmitted to the next generation. If he had, this would have provided an excellent support for Darwin's hypothesis that body parts shed gemmules that can influence the development of succeeding offspring.

No doubt all this should be regarded as so much negative evidence. However, Darwin dissuaded him from giving up his efforts. The new experiments were to insert the pollen cells or egg cells of a plant into a different variety of the plant. If adhesion takes place, the ovary might then be severed from its parent plant and left to develop to see which characteristics it takes on—host or graft. Darwin was advising him not to choose a plant ovary with a single ovule and not to bisect it after fertilization because he thought this would be quite a hopeless task. It would be better to operate on an ovary with many egg cells (ovules) but it still might be difficult to distinguish the effect of the union of two ovaries.

Darwin, now aged 72, was at the point of departing on the doorstep of Romanes' house when an attack seized him. He almost lost consciousness at the door. Romanes immediately asked him to come back into the house; it seemed that Darwin wanted to go home immediately because he felt so unwell. They offered to call him a cab, but he said he would pick one up at the end of the street. They watched him walk with difficulty down the street; after about 300 yards from the house, he staggered and caught hold of the park railings as if to prevent himself from falling. They were hastening to his assistance when after a few seconds he recovered and proceeded to find a cab for himself.

These seizures became more and more frequent and were associated with distressing sensations of exhaustion, faintness, and impending collapse. Several doctors saw him, including a Dr. Andrew Clark; Mr. Darwin's main treatment was left in the hands of Dr. Norman Moore of St. Bartholomew's Hospital. To no real avail. He had a further serious attack at Downe while sitting at dinner on April 18. He became giddy and staggered to the sofa where he fainted with his face down onto it. He again recovered slightly and seemed better until about midnight when he woke Emma saying *I have got a pain and I shall feel better or bear it better if you are awake.* He asked her to get a capsule of amyl nitrite from his study to improve the pains around his heart but by the time she returned he had fainted again. Brandy revived him and thinking that death was near he asked Emma to *tell all my children how good they have always been to me* and, as though to comfort her, *I am not in the least afraid of death.* During the night he had another attack and was found gray, cold, and breathing heavily. He died later that morning on April 19, 1882 in his 73rd year. His son said he would like his father's epitaph to read; *As for myself, I believe that I have acted rightly in steadily following and devoting my life to science.*

To the outside world Galton was grief-stricken. He wrote dutiful letters of condolence and to the family he wrote that he was absolutely sickened at the loss of Darwin. He wrote that he owed more to Darwin than to any man living or dead. He never entered Darwin's presence without feeling as a man in the presence of a beloved sovereign. Darwin was so wholly free of petty faults, so royally minded, and so helpful and sympathetic. It was a rare privilege to have known such a man who stood far above his contemporaries in the science of observation, and so on and so forth.

After the news of Darwin's death Galton went to the Royal Society to arrange that a request should be telegraphed to Darwin's family by the president in the name of the Royal Society asking if they would consent to an interment in Westminster Abbey. The funeral should be attended by a deputation from all the learned societies of Britain. He wrote to Lord

Aberdeen who fully agreed on behalf of the Geographical Society. Galton thought that Darwin should be laid by the side of Newton, as the two greatest Englishmen of science. He said the world seemed so empty to him now that Darwin had passed away; he reverenced and loved him so completely. There were similar effusive public letters in the *Pall Mall Gazette* and from the presidents of the many learned societies at home and abroad. He went on to organize a sub committee of the Royal Society to consider raising a permanent memorial or monument to his cousin Darwin, and to collect all the available memorabilia, such as pictures of his ship, the Beagle, to form a national depository at the Royal Society.

The burial in Westminster Abbey duly took place on April 26, 1882. Alfred Wallace, the naturalist who published the idea of natural selection at the same time as Darwin was the pallbearer, along with the biologist Thomas Huxley, the bulldog defender of Darwin's ideas in the public domain since Darwin was so frequently too unwell to fight on his own behalf. A prominent churchman was found, the Rev. Frederic Farrer to be the pallbearer, either with or without his ecclesiastical robes according to the wishes of the Darwin family. He was to preach the following Sunday on Darwin's work and said he wished to make such amends as he could for the reception formerly given by the church to Darwin's book on the *Origin of Species* (described in Epilogue 1); and Galton gave him some telling points to include in the sermon. Farrer said:

> *...Darwin will take his place, side by side with Linnaeus; with Newton, Pascal; and with Herschel and Faraday among those who have not only served humanity by their genius, but have also brightened its ideal by holy lives.*

Farrer was chosen because Darwin had been so impressed by Farrer's book on *Language and Languages* in which he presented an evolutionary interpretation of linguistics. In 1866, Darwin proposed Farrer for his election as Fellow to the Royal Society, which he indeed became.

Darwin's son William was sitting in the front row as the eldest child and chief mourner for his father when he felt a draught from behind. Because of the family's ingrained hypochondria he thought of his balding head and so might catch a nasty chill. He protected himself by putting his black gloves to balance on the top of his head; and he sat like this all through the service with the eyes of the nation on him.

Galton's biographer and most promising student Karl Pearson (1857–1936) wrote in his four volumes of *Life of Francis Galton* that Galton privately breathed a deep sigh of relief at Darwin's death. Galton told Pearson in so many words that it was difficult to measure what mental development an individual loses and what he gains by the death of a friend and mentor of such renown as Darwin. He personally felt he had gained by Darwin's death, however harsh and cruel this judgment might seem. Galton had been shackled by Darwin in the free range and development of his own ideas. He had been forced to retract the results of his experiments on heredity and to write books, publish articles, and give lectures about ideas based on Darwin's theories and not his own. That is the trouble with these damned dominant personalities, they can so easily stifle the imagination and creativity of those around them. Science could now advance by the funeral of one of its stars. Even Louisa Galton noticed that, although Darwin's death cast a temporary gloom over her husband, it was followed by his most productive decade. His statistical and eugenic studies became predominant and were of lasting interest. The year 1882 really marked a watershed for Galton's scientific work. The great man of Downe was no longer there to question minutely all the facts with only one idea in mind, nor to restrain Galton's free ranging imagination, nor to inhibit the opportunity to explore new avenues in his search for the truth.

For his part Romanes composed a touching ode:

> *I loved him with a strength of love*
> *Which man to man can only bear*
> *When one in station far above*
> *The rest of men, yet deigns to share*
> *A friendship true with those far down*
> *The ranks: as though a mighty king,*
> *Girt with his armies of renown*
> *Should call within his narrow ring*
> *Of counselors and chosen friends*
> *Some youth who scarce can understand*
> *How it began or how it ends,*
> *That he should grasp the monarch's hand.*

Sadly Darwin did not get the poet he deserved.

Huxley's epitaph was a little more pithy and sober:

> *None have fought better, and none have been more fortunate than Darwin. He found a great truth trodden underfoot, reviled by bigots, and ridiculed by all the world; he lived long enough to see it, chiefly by his own efforts indisputably established in science, inseparably incorporated with the common thoughts of men and only hated and feared by those who would revile but dare not. What shall a man desire more than this.*

Two years later in 1884 Mendel died from chronic kidney disease. The funeral was a quiet affair compared to Darwin's. The Czech composer, Leos Janacek (1854–1928) played a requiem on the organ for him. Mendel's obituary in the daily newspaper *Tagesbote* reported in conventional terms that *His death deprives the poor of a benefactor and mankind at large of a man of the noblest character ... one who was a warm friend, a promoter of the natural sciences and an exemplary priest.*

He was buried in the northeast corner of the city cemetery in a large plot reserved for the graves of members of the Augustinian monastery of Brünn. At the center of the plot is a large marble monument with an engraved line from Romans 14:8 *Whether we live or die we belong to the Lord.* Mendel's grave is off to the far right, his name almost obliterated by the growth of lichen and moss on the gravestone.

One can read what Mendel truly valued in his life from the objects he chose for his Coat of Arms divided into quadrants (see frontispiece): botany comes first with the picture of a biblical flower, the lily (not the Garden Pea); second comes the cross above a farm plough, perhaps referring to his farming ancestry; third are two clasped hands below the Sacred Heart, perhaps a tribute to his monastic community; and fourth are the scholarly letters of alpha:omega, the first and last letters of the Ionic Greek alphabet, which can also be a title for God (Revelation 1:8).

All Mendel's scientific papers, sermons, and notebooks were subsequently burned by the incoming Abbot. It has been said that this was done to draw a line under a taxation controversy in which Mendel had played a part. The dispute was with the local government to impose special taxes on religious institutions, which Mendel had opposed. It is likely that most people would consider Mendel's papers on heredity to be rubbish only and fit for incineration.

chapter twelve

The Grim Reaper revisits

I walk through the valley of the shadow of death.

Psalms 23:4

By June 1892, Romanes noticed a very worrying symptom. He had developed a blind spot in the vision of his right eye that gradually extended so much that he completely lost the upper half of his visual field. The blindness was so complete that if he looked about an inch below a light placed even at a short distance from him he was unable to perceive any luminosity.

He consulted Mr. Doyne, a well-known oculist in Oxford and told him that he had previously suffered from intermittent headaches; some of them were quite severe. Mr. Doyne found Romanes to have extensive fluid collection at the back of the eye on the retina and took a grave view of the prognosis. Romanes was told that the impairment would in all probability be permanent and so it would prevent all operative work on animals where any delicacy was required. Romanes went for a second opinion to an oculist in London who agreed with Mr. Doyne's assessment.

The next symptom to appear several months later was a temporary loss of speech that lasted several hours and then recovered completely. About 9 months later, he was struck down with a paralysis of the left side of the body. It did improve somewhat over the next few months. When he had a further episode of paralysis, he realized he was doomed at the early age of 45 years. Nobody told him in so many words, but he suspected that he had a brain tumor.[30]

He recalled that Mr. Darwin had said many times: *I am not in the least afraid to die*; and in saying this to his wife Romanes used exactly the same gesture that Darwin had used, probably unconsciously copying him. They both had wondered when their time should come whether they would be able to say the same? Romanes was 39 years younger than Darwin and felt quite bitter that his work was to be prematurely cut short so soon. He had five sons—the eldest not yet in his teens and the youngest still an infant. He had a pile of notebooks that no one else could use and heaps of experimental projects to complete. He said he felt like Job with everything being taken from him and very little prospects of a later reward, at least on this Earth. He slowly became a wreck of his former self. He entirely lost the power to write poetry and could not find the words to put them together properly. Galton never cared for his versifying, finding it too sweet and sentimental. He preferred the more robust and masculine verse of the poet Robert Browning. However, he felt very sorry for Romanes who showed a great deal of courage and stoicism in the face of desperate circumstances. Romanes was anxious and bitterly disappointed, realizing he was, as he said, entering a new land of pain and darkness, and was called on to join the great army of those who suffer. He was afraid to enter but it was now his turn—to endure. He played his part well and commanded the respect and admiration of his colleagues as well as the love of his friends. He patiently tried to resume his work. He continued writing his book on *Darwin and After Darwin* but eventually had to resort to dictation when he completely lost the use of his writing hand. Yet his faith in Christianity became stronger during this period; he often said—it is either Christianity or nothing for me.

Romanes told Galton that he wished he had been less ambitious for scientific applause and worldly success. These things are as nothing compared to the most precious things in life—faith and love. Worldly fame was out of the running altogether, there were so many setbacks and the prizes when attained seemed so insignificant. To him, even the love he bore for his wife and children was no compensation for a loss of faith. Fortunately, he retained this to the end. He said: *I have come to see that cleverness, success and achievements count for little in life; it is strength of character and integrity that is the important factor.* Romanes had perhaps attached an undue importance to intellect and social status in the early years of his career. The approach of the grim reaper had changed all this.

He managed to work for about two and a half hours each day, mainly dictating his book that progressed slowly. He found it very hard to bear that he was so much of an invalid and could do so little else. On September 4, 1893 Galton received a dictated letter from him: *the long and short of it is that I know I am dying. I have been gradually getting worse and worse ... nor shall I be sorry when it comes. Such being the case I should like to consult you about setting my papers and manuscripts for my book in order...* He sounded peaceful and resigned to the inevitable.

When descending into the valley of death, scientific disputes (is it gemmules or stirps?) seem to fade away into those elementary principles of good will, which bind all mankind together at times of crisis. As for courage, sooner or later death must come for us all; and Romanes courageously resolved to work as long as he possibly could. He still worried about gemmules and several months before his death was designing more experiments to test the hypothesis. He studied Professor Weismann's new book on *The Germ-Plasm: a theory of heredity* (1893). Two of Romanes's papers were read before the Royal Society, one of them describing the effects of light on plant growth. He replied to numerous letters from well-wishers and colleagues, including a long letter on adaptive evolution to Professor Henslow.

He repeated Darwin's maxim in telling Thomas Huxley in the house at St. John's Wood: *I am not in the least afraid to die.* Huxley responded rather tactlessly that the prospect of death filled him with an unspeakable horror. Huxley added:

> *Whether or not nature abhors a vacuum, I know that the soul of man does.*

Romanes' exhaustion seemed to grow on him week by week, so by the end of 1893 he was unable to walk and had very little hope left. His wife spent time reading to him—his favorite novels, poetry, and some history. In early 1894, he had an episode of prolonged vomiting and was kept in bed for 3 weeks. This passed and gave grounds for fresh hope. Perhaps he could live longer, although at a greatly reduced level if only his condition became stable. However, this was not to be. He came into his study at about midday on May 23, 1894 and asked about the book in which he was then interested, *Some Aspects of Theism* by Professor Knight of St. Andrew's University to be read aloud to him. Before reading he changed his mind and said he would prefer to lie down in his bedroom. On lying down he complained: *of feeling … something is not right… I can't breathe… God have mercy on me…* His wife knew he was going to die soon; and seeing his face suddenly grow white and pinched she guessed it was the end and became frightened. She rushed through all the rooms of the house calling out for help, so that the doorway of his bedroom was soon crowded with the cook, the housemaid, and the errand boy. The butler was giving orders to others who appeared not to understand what was happening. Romanes lapsed into a coma and never recovered consciousness again. He died in less than an hour. When the doctor arrived, he said it was a bad business, that Romanes was still so young, not yet 50. And with him died any further experimental work on Darwin's gemmule hypothesis.[31]

chapter thirteen

Mendel again—Galton's response

Life is short, art long, opportunity fleeting, experience fallacious and judgment difficult.

Hippocrates (c.460–c.377 BC)

Galton at 78 years felt privileged to have seen the old century out. Many of this friends and relations never reached 1900. His cousin Charles died in 1882. His beloved sister Adele died in 1883. Romanes never made to the century's end, dying in 1894 at the age of 46 years. And his wife Louisa passed peacefully away at the age of 75 years in 1897.

Galton was still active and strong but learning to live within the limitations of old age. His niece Eva had now taken charge of his household. She provided Galton with a delightful traveling companion; she was always cheerful, punctual, and interested in many things. Moreover, she always saw the good side of people, and never fussed about her health or got impatient or grumbled if they were kept waiting too long for their food or luggage while on their travels.

In many ways, Galton was happier now than he had ever been. He ate well, drank well, and had the freedom to travel where and when he liked. The only impairment was his increasing deafness. All his life he had been a relatively wealthy man and yet he never really appreciated his money. Now it gave him all the comforts that one could desire in declining age. He had more time to write letters and entertain friends. Sometimes he sat alone for hours. He was an old man and that is how old men live. There was time to read, to think, and prepare to write his memoirs. He still had a good health and a sound mind. He admitted to being old but he still maintained a stylish grip on himself and he still had the zest for travel. He had made tours almost every year since his marriage, usually abroad and often to watering places such as the spas of Austria and Southern France, and he resolved to continue these travels with Eva.

Eva and Frank spent a most exciting three months at Luxor in southern Egypt.[32] They stayed at the Hotel Karnack and made excursions for up to seven hours, including about 14 miles on a donkey ride to view the ruins. The wonders there were beyond belief. The massiveness and antiquity of the statues and temples left him breathless (which was a state he quite often found himself in for other reasons than wonder). They had a most interesting week's stay with an old acquaintance of Galton, Flinders

115

Petrie (1853–1942) at his archeological excavation about 100 miles away from Luxor. Petrie, a founding father of Egyptology (and archeology) was currently interested in the pre-Pharaonic people who had lived in Egypt before the great pyramids were built, that is, before about 4000 BC. He had found beautiful flint knives of a far earlier date, the most accomplished he had ever seen in workmanship and in artistic design.

In the spring, Eva and Frank spent a glorious 2 months in Rome. He had been there before in 1853 and again in 1886, although never really had the chance to look around the archeological parts of the city in detail. Eva was keen to draw and paint some of the more spectacular ruins. They stayed at the Hotel de l'Europe, very pleasantly situated for visits to the major archaeological sites. They had a delightful afternoon in the Forum and on the Palatine Hill visiting the recently excavated foundations and the vast palaces of the several Caesars. The Forum is such an extraordinary and exciting place, not a bombsite of ruins, but a site ruined by time and civilizations. Broken-up buildings, scraps of columns and pillars, enormous arches, blocks, and pediments scattered everywhere. They came to an inlaid marble staircase behind the Temple of Venus leading down into the ground. They descended these and at the tenth step, the staircase abruptly ended. They were tantalized to know where the steps would lead into the depths of the earth below; what large underground caverns and passages were laying beneath their feet.

It is then only a short walk from the Forum to the Palatine Hill that carved promontory set on one side of the Forum with a cliff face overlooking Rome. The cliff has been excavated into a series of caves and recessed caverns, cellars, and cryptoporticoes (subterranean passages); and beyond it is a rolling campagna of meadows dotted with olive trees with mounds of old stones lying around. The sun was shining, there was a wash of blue air over the distant vistas of Rome, and the whole day was as he wrote, *a veritable Indian summer to my life.*

Frank then had to return to England for a meeting on heredity to celebrate the discoveries of Mendel. They traveled home via Bologna, Milan, Cologne, and Brussels. They hoped to have spent Easter Sunday in Cologne but they had an enforced stay in Milan where Eva took to her bed for four days. Galton thought it was due to catching something from the possible sewer fumes emerging from the pipes in their hotel room in Rome, even though their rooms were quite high up. So, they spent Easter Sunday listening to music in the great Cathedral of Milan. They were more than satisfied with their tour of the south. They had never seen such greater beauty of sea, sky, and monuments than on this journey. They reached London on May 20, 1903.

Galton had, of course, heard of Gregor Mendel but at first he was not quite sure what he had done. The up-and-coming Cambridge scientist William Bateson (1861–1926) was organizing the London meeting, and had already written to Galton in 1900 suggesting that he should search out Mendel's papers to make sure he did not miss them. According to Bateson, it was one of the most remarkable investigations yet made on heredity. Galton came to appreciate this; he said his heart always warmed at the thought of Mendel, so painstaking, so unappreciated, and so scientifically solitary in his monastery[33] and was looking forward to hearing about Mendel's full details at the conference.

The Royal Horticultural Society of London was an unlikely venue for a scientific meeting, although many eminent scientists such as Galton were invited as guests to their conferences. The society was founded in 1804, the earliest of such societies in the world; however, it had very little scientific influence compared to the Royal Society or the British Association. It mainly consisted of a gentleman's club to provide elegant social events in London. Their flower show was always popular and attracted such notables as the queens of Sweden and Norway, the duchesses of Connaught and Devonshire, and Lord Cross. There was usually a lavish dinner in one of their halls to round the day off. To gain some scientific credibility, the society had begun to sponsor conferences on hybridization and plant breeding since 1899. Bateson was infiltrating the society for his own particular ends.[34] He had recently published a paper on the problems of heredity in the *Journal of the Royal Horticultural Society* to promulgate the ideas of Mendel. He had learned of Mendel's work by reading de Vries' new mutation theory. While traveling on a Great Eastern Railway train to a previous meeting of the Royal Horticultural Society in Liverpool, the 40 year old Bateson read about Mendel's paper from de Vries and immediately recognized its significance. Bateson's wife recorded that it was *as though there was a very long line to hoe that one suddenly finds a great part of it already done by someone else*. Bateson had already begun similar work and he rewrote part of his lecture on the train to Liverpool to include Mendel's results. He believed that Mendel confirmed his concept of discontinuous variation, as opposed to Darwin's view of continuous variation by gemmules.

He persuaded the Horticultural Society to pay for a translation into English of Mendel's original papers of 1866. Bateson incorporated this and his lecture into his book titled: *Mendel's Principles of Heredity: a Defense* published in 1902. Its aim was to promote and establish the rediscovery of Mendel's work.

What Huxley had done for Charles Darwin's theory of evolution, William Bateson was now doing for Gregor Mendel—defending him to the hilt against all adversaries, especially against Karl Pearson. Pearson was one of the most famous protégées of Galton and an ardent supporter of the idea of "blending inheritance." Bateson in 1903 was entering middle age but still only eking out his living from a fellowship at St. John's College, Cambridge and earning a little extra income as steward of the college. His wife in a later verbal portrait of her husband described him as:

> *Absent-minded, often mislaying his notebooks, forceps,*
> *scissors, pipe and glasses. Of his clothes he was as reckless*
> *as a schoolboy. He was capable of going up to London in*
> *old garden flannels, darned across the knee, or at the other*
> *extreme he might be found kneeling on the gritty garden*
> *path in a brand new town suit recording the growth of*
> *some batch of seedlings.*

He was clearly a suitable person to take on the role of an absentminded professor. Later he was to coin the term genetics and directed the "school of genetics" at Cambridge University in 1908. However, Bateson was yet to make his way in the world and had arranged several conferences, including one at the Royal Horticultural Society, hoping to disseminate and gain more converts to the Mendelian camp. Anyone who had contributed to the problems of heredity was to be invited.

The proceedings were quite straightforward. There was first a brief summary for newcomers to the subject of what Mendel had done. To recapitulate—Mendel had spent 8 years in his monastery at Brünn cultivating and hybridizing garden peas. He had studied and sorted as many as 10,000 plants, about 40,000 flowers, and classified more than 300,000 peas as reported by the Smithsonian Institution, United States. He patiently performed the process of hybridization by hand and counted which plant features from the parent plants turned up in later generations of hybrids. In the analysis of his results, he made two surprising observations:

1. Plant features that seem to be lost in one generation may crop up again a generation or two later in their exact original form, that is, by skipping a generation. For example, after the second cross, some peas of the subsequent generation of plants reverted back to their grandparental trait that was not seen in the original parents.
2. He found constant mathematical proportions recurring in his counts of the various inherited features, such as smooth or wrinkled peas that occurred in the second generation of hybrids. These rarer wrinkled peas appeared to turn up in the proportion of 1 in 3 with the normal smooth-coated peas; and this ratio was found for a number of other characteristics that he studied. He had discovered a mathematical constant in nature, in the form of a ratio that related to the inherited features of a plant. Discovery of such naturally occurring ratios is always an important event.

What did this ratio mean and where did it come from? What did it say about how these traits were inherited? Mendel created a model for inheritance that could explain how these ratios came about. He had lectured on his results and proposed a novel theory of *dominant and recessive inheritance* in February and March of 1865 before the Brünn Society for the Study of Natural Sciences. They were the foundation of the paper he published the following year in the official transactions of the society. These results are described fully in Chapter 16. This Brünn Society exchanged its publications with most of the academies of Europe, including the Royal and Linnaean Societies of England. As previously mentioned, Mendel ordered forty reprints from the journal editor to send to the most important scientists working in the field. Twelve of these have been traced, including two to Britain, so it was inexplicable that his paper should have passed unnoticed or been ignored for so long.

Unfortunately, Mendel's major scientific work ended three years later in 1868 when he was elected as abbot of the monastery. His eyesight was failing and he could not do the delicate work needed for cross-hybridization. He did not seem to think of handing over the project to a younger monk perhaps because his paper of 1866 met with such muted response from the scientific community at that time. So, he spent the remainder of his life as a leader for the monastery of St. Thomas until his death in 1884. From the ecclesiastical point of view, he became responsible for the spiritual care and development of several dozen monks and to act as a judge for issues of canon law among them. Some might argue these were far more important tasks than doing obscure experiments on peas. The monastery to the present day still has four monks but they are very elusive, and a curator of Mendel's museum in 2013 had not caught sight of one for the past 4 months.

William Bateson gave a marvelously clear account and defense of Mendel's discoveries, and started with: *An exact determination of the laws of heredity would probably work more to change man's outlook on the world and his mastery over nature than any other advance in biology that could be foreseen.* Sentiments that exactly coincided with those of Galton.

Toward the end of the lecture Galton for the first time really understood what Mendel had done. It was an extraordinary coincidence because Galton had been working on the inheritance of sweet peas, following a suggestion by Darwin, only about 5 years after Gregor Mendel had published his results on edible peas. It almost seemed too much of a coincidence and Galton wondered from where Darwin had got the idea. Was it from the paper Mendel had sent to Darwin or from any previous correspondence with Mendel or von Nageli in the 1860s? If so, none of the relevant letters or papers survived.

Gregor Mendel, however, did a much better job than what Galton and Darwin together had ever done. He arrived at some definite conclusions that replaced their own speculations about gemmules. It turned out that it was absolutely critical for his experiments to start with a pure line of edible pea, so that he could always reproduce the same phenomenon of crossing a tall plant with another tall plant to get seeds always producing tall plants. He studied all the other constantly inherited features in pure inbred lines of plants, including pea coat (whether smooth or wrinkled); shape of the ripe peapod (smooth or constricted around each pea in the pod); color of the unripe pod (green or yellow); and location of flowers (either restricted to the tip of the plant stem or arranged evenly along the whole stem). After he had done the complete analysis, he deduced a very plausible explanation that best accounted for his observations of the constant numerical proportions of the different characteristics that he found after successive crosses. The numerical proportions in the second generation cross, in particular, gave the very striking ratio of 1:3.

There was some difficulty in following the arguments; however, by the end of the talk the penny had certainly dropped with many people. Galton suddenly had this very sharp picture of how inheritance worked. He had found Mendelian ratios himself purely on theoretical grounds; but Darwin had given it a short shrift. In their joint experiments they perhaps should have limited themselves to what they could actually observe and measure, instead of trying to force the observations to fit Darwin's hypothesis involving abstract things such as gemmules that they could never detect.

Bateson's lecture ended with these inspiring words: *That we are in the presence of a new principle of the highest importance is I think manifest. To what further conclusions it may lead us cannot yet be foretold.*

How could Galton have been so blinkered as to have missed this himself? Especially, when he was so close to making the same observations. To quote a letter of Galton's to Darwin in 1875 after Darwin had criticized his modified gemmule (or what Galton called germ) theory of inheritance. At the time, Galton was considering the likely inherited features of crossing plants with white-colored flowers to plants of a black flower variety. He was attempting to predict flower colors of the hybrid plants if the germs (to use his terminology) were organized on stirps (that is like corn seeds on a cob) and not flowing freely from all parts of the body. Galton's letter reads:

> *If there were 2 gemmules only, each of which might determine white or black colored petals, then in a large number of crosses, one quarter would always be quite white, one quarter black, and one half would be gray.*

These were some of the mathematical ratios that one would expect using the binomial theorem with which he was perfectly familiar. It was a cornerstone of Mendel's hypothesis that inherited factors come in pairs: one from the male, and one from the female. However, Galton had not thought of the idea of dominant and recessive factors that would have then given him the Mendelian ratios of 1:3 and not 1:2:1 as he predicted from his flower crosses (codominant heredity). Then he had done the experiments on sweet peas but measured blending features of pea size and weight, and not the all-or-none features such as pollen grain shape as either round or oblong.

Furthermore, he had published studies on the inheritance of coat colors of dogs, particularly the Basset hound. He had been given access to a large pedigree stock of hounds owned by Sir Everett Millais, eldest son of the pre-Raphaelite painter John Everett Millais. The Bassets are dwarf bloodhounds with only two alternate varieties of coat color, inherited either as white with blotches ranging from brown to yellow, or alternatively with black blotches. The pack contained 817 hounds whose parents were available, 567 hounds where the coat color of the two parents and four grandparents were known, and finally, there were 188 hounds in whom the coat colors of all eight great grandparents were also known. The results showed that the coat colors of the dogs tended to "breed true" and there was little evidence for the appearance of intermediate or mixed fur colors. Galton called this particulate inheritance to distinguish it from blending inheritance and from these data deduced his ancestral law of inheritance. In some ways, Galton's mental energy and versatility were his own worst enemy. He started off on too many projects all in different directions and his ideas never gelled into a consistent theory. Whereas Mendel pursued just one project with a correct theoretical basis that provided the foundation of genetics for the future.

Galton generously paid tribute to Mendel in his *Memories of My Life* published in 1908 well after Mendel's views had become generally accepted. He wrote *His [...Mendel's..] careful and long continued experiments show how much can be performed who like him and Charles Darwin never or hardly ever leave their homes, and again how much might be done in a fixed laboratory after a uniform tradition of work has been established.* He went on that Mendel had clearly shown that there were such things as alternative discrete characters of inheritance of equal potency. Galton here appears not to have understood the importance of Mendel's idea of Dominant and Recessive inheritance. Galton then wondered if such inherited features are due to simple inherited molecular characters or to many of them correlated together (is he still harking back to gemmules?). This project on heredity shows how long and with what difficulty a new scientific truth takes to get established; and one has to wait for Galton's death before the gemmule hypothesis of heredity dies out all together.

Darwin rejected Galton's ideas because they did not fit in with his idea of multiple gemmules originating from all the bodily parts. The tests to demonstrate these gemmules would be the grafting experiments; doing crosses and backcrosses of plants or animals would be next to useless according to Darwin. Therefore, Galton never got to test his own ideas by the experimental method. He thought he was in possession of an idea that was superior to Darwin's; despite all his arguments, he probably lost confidence when Darwin told him by letter that his views were curious and muddled, and that the obscurity was not in his (Darwin's) head but in Galton's. He was being disregarded simply because his voice was not powerful enough to overcome the authority of Charles Darwin. Experts tend to exaggerate the importance of their own ideas, the longer they hold them and the more difficulties they have encountered to establish them. Whenever a scientist upholds his ideas with a blind faith, then one should seriously doubt the value of that idea. Indeed, great men can often teach us by their errors as much as by their discoveries. However, error of opinion is only tolerable where reason is left free in others to combat it. The gemmule model of inheritance would have undoubtedly have remained a vague concept even if Mendel had not formulated and tested his own remarkable theory of heredity.

Darwin appeared to be less receptive to new ideas that were not his own. For example, he complained bitterly about the wretchedness of Lamarck's book *Philosophie Zoologique*, especially if his own ideas were ever identified with Lamarckism. Yet Lamarck's book clearly presented a scheme for the evolution of species, graded by increasing complexity, and leading to the origin of man. What Lamarck got wrong was how evolution worked, postulating that bodily characters acquired through interaction with the environment could be inherited. However, Darwin's own views about heredity based on false assumptions such as the existence of gemmules doom even the best pieces of experimental work such as Galton's. There was no deficiency of logic in Darwin's ideas, although this is often the case with some scientists. If only they had kept their attention fixed on the entire problem and not just on one section of it that could be explained by gemmules.

Galton would have been dismayed if he had learned that a reprint of Mendel's paper might have been received by Darwin. Of all the biologists working between 1866 and 1900 in England, Galton was probably the only one who would have appreciated the implications of the paper with his interest in discontinuous variation and his mathematical turn of mind. Galton had even devised a binomial machine (now called the Galton Board) that gave ratios of distribution of ball bearings or lead shot when poured into a flask with appropriate flow dividers that matched the 1:2:1 ratios Mendel actually found in some of his hybridization experiments. Both were based on a binomial expansion to the power of 2, so Galton was quite familiar with the mathematics that Mendel had employed. The fact that Darwin may have read Mendel's article and not bothered to inform

Galton anything about it except that he should perhaps do experiments on the heredity of the sweet pea must have been deeply disappointing. Darwin may have been put off by the mathematics in Mendel's work. Darwin wrote in his autobiography that he deeply regretted that he did not proceed far enough in his studies at least to understand something of the basic principles of mathematics, but he believed that he would only have progressed to a very low grade.

Mendel's observations and his concepts explained so many things that had puzzled both Darwin and Galton before. It explained an all-or-none inheritance because its basis rested on one particulate factor being passed on from each parent to the offspring and not a variable stream of particles coming from different bodily organs as proposed by Darwin; it explained the phenomenon of blending inheritance if one supposed that several elements of heredity from each parent were required together to produce the inherited feature; it explained "reversion" in which a trait in the offspring resembles as that found in the grandparents but not the parents; and finally it explained the origin of a "sport," that is, the sudden appearance of a new feature in a family such as a cleft palate or malformed ears. Here, one could imagine that an element of heredity undergoes a sudden change in character, perhaps chemically, to produce an entirely new element that is responsible for the new feature that can go on to be inherited by successive generations. Mendel's "Laws of Heredity" made it possible to foretell in statistical terms the results of observations that would be made if different varieties of the edible garden pea were crossed. Mendel had failed to qualify as a teacher but he had successfully tackled a central problem of biology and provided a mathematical formulation of the natural laws underlying the basis of heredity. Mendel inverts the old adage that *those that can, do—those that cannot, teach*; now becomes for Mendel *those that cannot teach—instead can do.*

Galton never set out to show that Darwin's hypothesis was wrong. However, Mendel's work now strongly suggested that this was so. Galton wondered if Darwin would ever have admitted it. Darwin might have argued that because the model works for plants it does not necessarily mean that it will work for animals or man. Or perhaps the effects of aging and long-standing ill health on Darwin may have taken its toll and he may have been just too fatigued to take an interest in new viewpoints.

Although Galton and Darwin never demonstrated the existence of multiple gemmules, the idea had some value in pointing the way through the existing confusion of the subject to some definite experiments, such as blood transfusions in rabbits. It gave them a direction and a line of attack to work on; but perhaps they should have changed course sooner than they did. They sought for clear-cut conceptions, precision of expression, and a definite terminology. They postulated that the information for the transmission of inherited features was carried by particles of some kind, and that these particles (the gemmules) were capable of self-replication, as are the genes. They introduced a scientific method into the study of inheritance that had not been done before Mendel. They had an original idea from which they devised a hypothesis that could be tested experimentally. They went on to design experiments to test the hypothesis using rabbits as an experimental model. Measurements were made, similar to Mendel, by the simplest technology of the day, which is by counting the number of rabbits that were altered. They analyzed the data and came to conclusions that were published in reputable scientific journals. Their leading idea, however, was wrong, which they should have eventually realized. Their experimental design should have switched from grafting-type experiments to cross-breeding experiments, which in fact, Galton wanted to and almost succeeded in doing. Everything becomes clear in retrospect.

chapter fourteen

Are Mendel's ideas true?[35]

> …..*On a huge hill,*
> *Cragged and steep, Truth stands and he that will*
> *Reach her, about and about must go.*

On Religion (Satire111). John Donne (1572–1631)

Archibald Garrod (1857–1936)[36]

Galton was a great friend of the Garrod family and for a short time they had been his neighbors. He had known one of the boys, Archibald Garrod from childhood. In the 1890s, he was working in London at St. Bartholomew's Hospital in the same field as Galton—namely human heredity. Galton probably first heard about Archie Garrod's work (from childhood the boy hated to be called Archibald) from Dr. Norman Moore who was working at the same hospital. This was the same Dr. Moore who had looked after Darwin at the time of his last illness in 1882 and had met Galton several times during that time.

Archie Garrod published the major results of his studies in the *Lancet* in 1902. These demonstrated very suggestively that Mendel's model of inheritance worked for man too.

Archie was the youngest son of Sir Alfred Baring Garrod and Galton had visited their house quite often when the children were young. Sir Alfred was an able and talented doctor. He had a romantic story to account for the name of Baring. His great grandfather (a tenant farmer of modest means living in Suffolk) had once rescued a member of the Baring merchant banking family from the clutches of a dangerous highwayman. As a reward he was promised that any of his future sons who were given the name of Baring should be provided with sufficient finance to complete their education. Sir Alfred was, in fact, quite able to make his own way in the world without this aid. His first claim to fame was the important discovery that patients suffering from gout had raised levels of a substance called uric acid in their blood. It turned out to be a good diagnostic test to distinguish the disease from other joint disorders such as rheumatoid arthritis. He went on to specialize in rheumatic disease and there is a street in Aix-les- Bains in Southern France named after him, *Rue Sir Alfred Garrod*. This is because Sir Alfred stressed the value of taking the waters there for the treatment of gouty arthritis, and his opinion attracted more than 1,200 new patients to the Spa Town in 1 year. The town council considered that Garrod was responsible for making a sojourn at Aix a standard treatment for gout. Sir Alfred was later appointed as a professor to Kings College in the Strand but gave up the post at the age of 55 years to concentrate on his lucrative private practice. He was appointed as a physician to the queen in 1888 and then was able to raise his consultation fee in Harley Street from one to two guineas. He died a wealthy man leaving an estate of £84,551, as well as legacies to his coachman, butler, cook, and to his other servants who had been with the household for more than 5 years.

The Garrods had six children: four boys and two girls. Archie was the fifth child. Once he came with his sister Edith to Galton's house in Rutland Gate, so that Galton could take their fingerprints. They were such a distinguished family that Galton wanted to add them to his collection of notable pedigrees. He forgot to provide them with any means of washing off the printer's ink, so they had to go about London all day with sticky black hands. The eldest son, Alfred Henry, was a talented zoologist and was elected at the age of 30 years to the Royal Society; tragically, he died of tuberculosis 3 years later with a massive lung hemorrhage. Another son, Herbert Baring, was a barrister and an excellent classical scholar winning the coveted Newdigate Prize at Oxford University. Of the two daughters, Helen died of tuberculosis at the age of 32 years and the other daughter, Edith, kept house for her parents.

Galton had not followed Archie's career closely and it was a surprise to cross paths with him again in such an unexpected way. He was a nice-looking and well-mannered young lad when Galton knew him. He went to a public school, Marlborough College, and then to Christchurch Oxford. He qualified in medicine in 1885 at St. Bartholomew's Hospital in London (Figure 14.1). This is one of the oldest and most distinguished hospitals in Europe, founded in 1123. Its most famous staff member was William Harvey (1578–1657) who was a contemporary of Galileo (1564–1642), and both worked at the University of Padua in the early 1600s, the former studying medicine, the latter teaching mathematics. Galileo's fellow student William Harvey wrote in a letter of 1657:

> *Nor is there any surer route to the proper practice of medicine than if someone gives his mind over to discerning the customary laws of Nature through the careful investigation of diseases that are of rare occurrence.*

Whether Archie Garrod read this or not in the 1890s he started an enquiry into rare urinary pigments that he thought may be inherited from the parents—what turns the urine red (porphyria), or yellow (urobilin), or green (dye consumption), or black. It was the study of black urine that

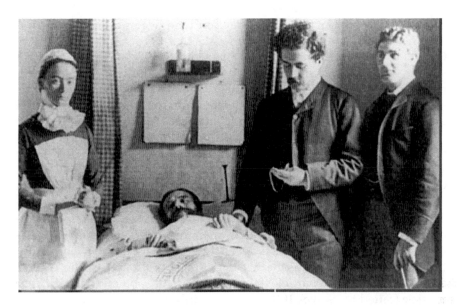

Figure 14.1 Archibald Garrod doing a ward round at St. Bartholomew's Hospital. He was the first to show that Mendelian rules apply to humans, and that a defective Mendelian *Elemente* causes an enzyme defect; in the case of alkaptonuria a defect in homogentisic acid oxidase. (Courtesy of Simon Garrod.)

had turned up the best results for him. The feature that Dr. Garrod was studying in his patients was a mild condition that turned the urine of the victim black after standing at room temperature for several hours. Dr. Garrod believed it was not a disease but only a harmless peculiar trait of the body's chemistry; in fact, over the long term it can cause quite serious joint and kidney damage. It was certainly not due to an infection, as previously thought, since it can start from birth with nappies of the infant turning black. It is called alkaptonuria, or more familiarly "black urine disease." He knew that the urine turns black because it invariably contains a chemical substance called homogentisic acid. This chemical spontaneously polymerizes in the urine to produce black substances akin to the black pigments of the skin, the family of melanins that produce tanning of the skin after exposure to sunlight.

Garrod studied the distribution of the condition in affected families. He found parents of affected children are usually closely related to themselves, usually first cousins; however, the parents do not show any signs of the trait. He collected the results of all the offspring of forty such families that had been reported in the world literature and tabulated how many affected children were to be found with the condition. The ratios almost exactly corresponded with those predicted by the laws of heredity as proposed by Mendel in his studies of hybrids of the garden pea.

This was pointed out to him by Bateson who began a correspondence after hearing about the preliminary results that were published earlier by Garrod in 1899. After that they became good friends, both coming from similar social backgrounds of wealthy Victorian families headed by eminent fathers and both getting first class degrees at university. Bateson was 4 years older than Garrod and stimulated him to continue the work and encouraged him to think of the results in Mendelian terms.

The actual numbers of individuals with "black urine disease" from the first six families of first cousin marriages that Garrod studied were: 12 affected people compared to 36 unaffected, giving a ratio of exactly 1:3. Garrod adopted the model proposed by Mendel that the hereditary unit consists of two particles (one from each parent) that determines the condition. One particle (the word "gene" had not been coined yet) is for the normal metabolism of homogentisic acid and anyone carrying this particle is unaffected. The other particle is defective and produces the "black urine" trait, in which homogentisic acid accumulates in the urine. The parents each have one normal particle (he called dominant) and one "black urine" particle (called recessive). If these two are inherited together the person's urine is normal. However, on mating with a partner who also carries the "black urine" particle, there is a random assortment to their offspring and one unlucky child in three, by chance, will inherit two "black urine" particles. It is they who will, therefore, excrete homogentisic acid in their urine and display the condition of black urine disease (alkaptonuria). Marriages of first cousins are most likely to reveal the condition because first cousins are more likely to bear the same "black urine" particles in their Mendelian units of heredity. These findings were first published by the Evolution Committee of the Royal Society during 1901 under the aegis of Bateson. In 1902, Garrod published his results in the *Lancet* that demonstrated very suggestively that Mendel's model of inheritance works for man.

Garrod went further than Mendel by suggesting that the accumulation of homogentisic acid in the urine might be due to a block in the normal pathways of disposal of this chemical. He fed homogentisic acid by mouth to patients and it was fully excreted in the urine. Since these chemical reactions in the body are under the control of enzymes it is possible that the person with the "black urine" particle is making a defective enzyme that cannot break down homogentisic acid, and so it is excreted. An enzyme is a special tool in the cell usually made of protein that facilitates the factory of chemical reactions that goes on in living cells. This gives Mendel's unit of heredity a biochemical reality. The heredity units are actually making proteins such as enzymes to regulate the chemical reactions going on in cells.

However, one swallow does not make a summer. Were there any other examples that might support this idea? No idea can be accepted as true until it has been tried in the fire of further experimentation.

In fact, Garrod had collected the same information for albino families. This inherited condition may be looked on as chemical in its basis, due to a failure to produce the black pigments of the melanin group in the skin. When he studied the incidence of albinism in such first cousin marriages he found the same numerical ratios, about one affected individual in three unaffected individuals, as he found for black urine disease (alkaptonuria). He had studied a third condition cystinuria, which unlike alkaptonuria, is a distinctly harmful condition because the chemical substance crystallizes out in the urine to form kidney stones. Again the pattern of inheritance in affected families fitted the Mendelian model. So, Garrod readily agreed that the laws of heredity as discovered by Mendel offered the best explanation for his own observations.

Clearly these studies had far-reaching implications for human inheritance. The work would have progressed much further, had not the hospital been particularly mean-spirited toward Garrod. They had not yet promoted him to the consultant staff of St. Bartholomew's Hospital, being only an assistant doctor and a chemical pathologist there. He had applied unsuccessfully for the post of staff physician in 1887, again in 1893, and twice in 1895. When Dr. Moore (the doctor who had looked after Darwin in his terminal illness) was promoted to the rank of a full physician, it left a vacancy and this time Garrod was appointed as physician in 1912 at the age of 55 years. He had begged "His Royal Highness, the President, the Treasurer, and Governors of St. Bartholomew's Hospital to offer himself as a candidate for the post of Assistant Physician to your Hospital." He was supported by 29 testimonials. Each letter could not have been more enthusiastic about his abilities and character; and all the writers showed their total embarrassment that Garrod had not been elected to the hospital staff sooner. It was disgraceful for a man with such analytical abilities and imagination to be blocked like that for so long.

Most of the doctors at St. Bartholomew's Hospital at the time could think no further than the size of the fee they were likely to collect from their next private patient. Garrod had to earn his living by attending another hospital for sick children at Great Ormond Street. In some ways, it had been a blessing in disguise for him. It had given him time to "play around" with his urine specimens, as his colleagues liked to call it. During this time he had made some fundamental progress in the study of heredity. When the two Nobel Prize winners, Beadle and Tatum, were giving their acceptance lecture on their one gene–one enzyme hypothesis in Stockholm 45 years later in 1958 they acknowledged that all they had done was to redemonstrate Garrod's discoveries in a different experimental system, the bread mold (*Neurospora*), a much easier model to work with than humans.

Garrod deserved heartfelt congratulations on his groundbreaking work. If only people would stop finding faults in the personal characteristics of colleagues. It does not matter at all whether someone is a professor of medicine, or a hospital physician, or a chemical pathologist, if he possesses the imagination and the desire for original observation and experimentation. It is not of decisive significance whether he has a vast amount of data or only a modest amount at hand. If he is in the position to ask the right questions and to find the right methods for answering them, this is all that should be required of him.

Galton's curiosity would certainly be aroused when Sir Alfred's young son Archie was now working in the field of human heredity, and in the space of a few months Galton had probably seen the results of observations and experiments from entirely different sources supporting the truth of Mendel's hypothesis. Corroboration of Mendel's works from such an unlikely a place as St. Bartholomew's Hospital on a completely different experimental organism, namely humans, made Mendel's theory even more likely to be true. The 1:3 ratios remained constant despite the wide range of plant and human characteristics that were being studied. Admittedly Garrod had studied fewer instances than Mendel but a few observations on a completely different experimental setup can be far more informative than more and more experiments with the edible pea or other plants. A large number of experiments may not be always as useful for verification as smaller numbers subjected to different conditions. Of course, these two studies did not prove Mendel's model to be true, but all the evidence made it more likely than any rival theory because it explained more of the facts over a wider range of subject matters.

Three botanists[37]

Were there any other reports in the scientific literature of the time to support Mendel's and Garrod's observations? The best libraries to search at the time in London would be the British Library housed at the British Museum or the library at the Royal Society of Medicine in Wimpole Street, London. Garrod would have to go through the index for each yearly volume looking for key words such as heredity, inheritance, hybrids, gemmules, and Mendel. It would be quite a daunting task just to go through the English journals such as *Nature* and the *Proceedings of the Royal Society*. He would have to include all the major European journals such as *Comptes Rendues* and *Arch. Ges. Physiol.*

One article with a promising title was published by a Dutch Botanist, Hugo de Vries (1845–1935), in *Comptes Rendues* (Paris) 1900. The title read *La loi de disjonction des Hybrides,* or in English translation: "Concerning the law of segregation of hybrids." de Vries had been studying the inherited features of the evening primrose (*Oenothera lamarckiana*). This is a native plant of the United States and was brought to Europe in the seventeenth and eighteenth century as a garden plant, possibly at first for the Jardin des Plantes at Paris. Since then, the plant had run wild in Holland and other countries. de Vries worked in Rotterdam and had grown this plant in large quantities from seeds over several years and found that a few individual plants were of such a different appearance in foliage, mode of growth, size, and so on as to appear like distinct species. He called these new forms sudden leaps (or "mutations"), and believed that new species of plant could arise by jumps in their structures, rather than the slower gradual process of continuous variation and selection as proposed by Darwin. To explore this idea he had started a systematic program of plant breeding in 1892 and during the next 8 years he had found that on crossbreeding several hundreds of supposedly pure strains of the primrose, the inheritance of particular features occurred in the same 3:1 ratios discovered by Mendel in his scientific paper of 1866.

There was another paper in the *Deutschen Botanischen Gessellschaft* of 1900 titled (in English translation): "Mendel's laws concerning the behavior of progeny of varietal hybrids." The author was a professor from Tübingen in Germany, Carl Correns (1864–1933), who had started the same experiments as Mendel, crossbreeding pure strains of the edible pea, and then moved on to study several varieties of maize. Only later when searching the scientific literature had he come across Mendel's paper of 1866. In addition, Correns had found that the same 3:1 ratios hold good for the variation of inherited features such as flower color, shape of pea, or color of seed coat, as Mendel had described.

A third paper was published in 1900 by Erich von Tschermak (1871–1962) in a top Austrian journal. It was a very long and detailed account of researches into the crossing of a number of varieties of the edible pea *Pisum sativum* (*P.s. quadratum, P.s. saccharatum and P.s. umbellatum*). The results confirmed the main facts as published by Mendel beyond any possibility of doubt. In fact, his results showed even closer to 3:1 ratios than Mendel had done (he found the ratio to be 3.008:1). The three authors taken together abundantly supported the general applicability of Mendel's laws of inheritance at least to the plants that had been studied.

Tschermak wrote that *The simultaneous discovery of Mendel by Correns, de Vries and myself appears to me especially gratifying...* Not so for Correns and de Vries who started a dispute about priorities. The history of science is littered with many priority disputes, the one between Leibniz and Newton about who discovered the calculus being a famous example; a less savory example was between Owen and Darwin about who discovered natural selection. The present one seems to have been another petty-minded affair. The gist of it was that Correns claimed to have discovered the 1:3 ratio in 1899, months before he had even read Mendel's paper; it was an entirely original observation for him and not just a confirmatory paper for Mendel's work. So he could claim priority. Correns after reading de Vries' paper of 1900 suspected that de Vries too wanted to hide Mendel's earlier discovery of the segregation ratio to claim originality for himself. Correns realized he had lost the priority for the discovery of the ratio and quickly published a second paper in 1900 pointing out that Mendel had already discovered it in 1865. It then appeared that de Vries did not properly understand the significance (or just lost interest) in Mendel's theory and went back to his own intracellular theory of "mutation" and evolution. He claimed that Mendel's data only applied to the special case of the garden pea. Later on, de Vries refused to sign a petition calling for the construction of a memorial to Mendel in Brünn and even rejected an invitation to attend the 1922 celebrations of Mendel's work. As he explained in a letter to his friend Friedrich Went in September 1922:

> *To my regret I cannot accede to your request. I just do not understand why the academy would be so interested in the Mendel's celebrations. The honoring of Mendel is a matter of fashion which everyone, also those without much understanding, can share; this fashion is bound to disappear. The celebration in Brünn is nationalistic and anti-English, directed especially against Darwin and thus unsympathetic to my mind, but therefore also very popular.*

He could have written that he was perhaps eaten up with envy and jealousy at the widespread acceptance of Mendel's ideas on heredity and the total eclipse of his own. Most scientists never let their minds be so overwhelmed by feelings of indignation and envy as to prevent them from giving a fair hearing to proposals made on behalf of their subject even if it violently disagrees with their own views. All men can make mistakes and are usually willing to acknowledge their errors if clearly demonstrated by others to be so, however galling it may be if the work is done by a younger and less experienced scientist.

Galton now realized that he was in the presence of a new principle of the highest importance. To what further conclusions these laws could lead regarding the nature of heredity was a tantalizing thought. It clearly showed that future experiments should be designed in such a way as to bring this theory of heredity to even more rigorous tests. Each experiment should be designed to pose a particular question and the range of observations restricted to the inheritance of one or at most a small group of characters. The fur color of rabbits might have been too complicated—it might not be based on a simple form of inheritance.

Four lines of research involving different experimental situations (humans, evening primrose, maize, and edible pea) fully confirmed Mendel's work done 35 years before and mathematical rules appear to govern the biological science of heredity. It meant that Mendel's model for inheritance had a widespread applicability and virtually proved it, or something close to it, to be correct in its basic formulation. It finally buried Darwin's idea of a multiplicity of gemmules from all the body organs because that theory could never give rise to such exact numerical proportions in the inheritance of any particular bodily feature in successive generations. The major issue was that neither Darwin nor Galton managed to arrive at a definitive elucidation of gemmules that could delimit it clearly from other concepts about heredity, and so never yielded to a systematic exploration of the subject matter.

Alfred Russell Wallace, a friend and colleague of Darwin, was another eminent British biologist who failed to recognize the importance of Mendel's work. He died in 1913 and so was well aware of the rediscovery of Mendel's ideas around 1900 by Garrod and the botanists. He published his opinions in *Contemporary Reviews* (1908) in an article titled the "Present Position of Darwinism." After giving a brief biographical description of Mendel he went on to harangue the work. He agreed that Mendel showed that certain pairs of characters in peas are inherited as "all or none" showing no signs of blending inheritance. He mentions the same phenomenon in plants, mice, rabbits, poultry, and so on. In each case Wallace argued that the feature studied is either a "sport," or of doubtful value for natural selection. Thus, in mice one of the parents was an albino; in rabbits one parent was an albino or long-haired; in poultry one parent had an abnormal comb. So, Mendelism is only concerned with the inheritance of abnormal forms that rarely or never occur under natural conditions, and so can have nothing to do with the origin or modification species in the wild. This is a false thesis if ever there was one. Technically, it is called the fallacy of initial predication. This means that just because the variants that Wallace picks out are albinism in mice and in rabbits or abnormal combs in poultry it does not mean that all the other Mendelian variants in further studies are going to be abnormal.

Indeed Mendel's selected features of the garden pea are all of natural occurrence.

According to Wallace, Mendel's concepts are not novel either, because most of the facts were known to Darwin. The fundamental fact that certain characteristics do not blend when hybridized was not only well known to Darwin but also carefully discussed by him in his book *Animals and Plants under Domestication*. Thus, when gray and white mice are paired, the young are not piebald nor of an intermediate gray tint but are either pure white or the ordinary gray variety; and the same occurred when white and common turtle doves were crossed. Darwin goes on to discuss many other examples of discontinuous inheritance in animals and plants. The reason why Darwin probably did not develop this research further to detect the numerical laws in successive generations similar to what Mendel had done, was that he felt quite satisfied from the large mass

of facts he had accumulated during 20 years of research that hybridization or the intercrossing of very distinct varieties had no place whatever
in the natural process of species formation. So, it seemed to Wallace, as a
supporter of his dead colleague, that Mendelism was only of the slightest
importance in the scheme of the evolution of life. Mendelism deals with
abnormalities whether called variants, mutations, or sports that would
play very little part in the transformation of species and could even lead
to their extinction.

chapter fifteen

Darwin and Mendel: The show down[38]

For now we see through a glass darkly; but then face to face: now I know in part; but then shall I know ...

1 Corinthians 13:12

By 1903, Galton was an old man of 81; but he was content with this. He had outlived all his friends and received more than his fair share of honors. He had been elected to two learned societies; and awarded three gold medals (and one in silver from the French Geographical Society). He had been awarded two honorary degrees: one from Oxford and one from Cambridge. He had been on terms of intimacy with many of the best scholars in Europe for the past 50 years. Lately, he had been awarded a Knighthood but it had come too late for him to enjoy it. He felt too weak and was too frail to attend Buckingham Palace to receive it. He asked to be excused. Apparently, it could be delivered by the postman. He led a peaceful and quiet marriage with Louisa, but without children he could hardly call it a family marriage as he had known when a child. He may well have felt mortified that his desire to have children had met with such a muted response from Louisa.

His brain was now not as good as it used to be. His memory was weaker. He could not remember the simplest of things, such as the names of his friends or colleagues of the 1880s. He could picture their faces but their names hovered on the tip of his tongue and he just could not recall them. Sometimes their names would pop up at the most unexpected of times, when he got into bed or when he was shaving in the mornings. He did manage to give a lecture at the London School of Economics in 1904. He was 82 then, and managed to stumble his way through the talk. In the past, no argument, no entertainment, and no sport had ever given him so much enjoyment as lecturing. Only when lecturing had he really been able to let himself go. Not even his exploration of Africa had left him as voluptuously exhausted as he felt after giving a lecture. This time, however, the feeling of exhaustion preceded the start of the lecture; and he thought the audience came more to see if an 82-year-old could hold his thoughts and concentration together rather than to learn anything from the lecture. To paraphrase a saying of Dr. Johnson: *that an old man of eighty giving a good lecture is like a dog walking on his hind legs. It is not done well; but you are surprised to find it done at all.* When he was invited 3 years later to give the Herbert Spencer Lecture at Oxford, he had the wits enough to write the lecture out, but not the strength to give it. It had to be read for him by one of his colleagues. It was clear that he was going to die in harness, as they say, still believing in things such as science, reason, and progress.

He had also been fortunate in his choice of students at University College London. He had found one to carry on his interests. This was Karl Pearson and Galton was immensely proud of him. Pearson had imagination, inventiveness and above all a taste for hard work. He had qualified as a mathematician at Cambridge; and then went on to practice law at the Inner Temple in London. During this time, he published two novels titled *The New Werther* and *The Trinity*. Then he obtained a professorship in applied mathematics at the University College London as well as being elected as a Fellow of the Royal Society—another true polymath similar to Galton. Pearson maintained Galton's interest in the field of heredity by lively discussions and constant intellectual exercise. Galton treated him more like his son, and he treated Galton as a foster father. He was so attentive, he kept Galton informed on all that was going on; he asked Galton's advice about important matters; and he had more than a touch of reverence for Galton as Galton had had for Darwin. Galton should have warned the poor fellow off—although Galton did enjoy and bask in this attention and adulation. One thing that upset Galton was that Pearson had taken a strong exception to Mendel's work, and spent too much of his time fighting with supporters of Mendel's ideas, particularly with William Bateson. Pearson unreservedly supported Darwin's and Galton's blending views of inheritance and he wrote of Galton's ancestral law of heredity of 1897 that: *it is a complete solution, at any rate to a first approximation, of the whole problem of heredity.* Galton, similar to Mendel, had tried to express the laws of heredity by mathematics, except he did not choose the binomial theorem; instead, he used the idea of a convergent series. This is a set of things ranged in a particular order. Consider the sequence of U.S. presidents of the twentieth century arranged in the order of their first tenure of office. The sequence commences with T. Roosevelt and ends with George W. Bush Jr.

In the same way Galton considered the sequence of ancestors of a child who might contribute to his or her heredity; namely, the parents, grandparents, great grandparents, and so on. Galton's law then stated that: the two parents contribute between them on the average one-half (1/2) of the total heritage of the offspring; the four grandparents one quarter (1/4); the eight great grandparents one eighth (1/8); and so on. Then the sum of the ancestral contributions to inheritance is expressed by the series:

$$F(x = 1) = \left[(0.5) + (0.5)^2 + (0.5)^3 \ldots (0.5)^n \right]$$

where $F(x)$ is the summed inheritance of the child.

This is a geometric series and its sum approximates at its limit to the value of 1; but of course, each of the terms is not independent of each other as the equation might imply. All the contributions of the grandparents and great grandparents and so on are transmitted through the first term of the parents. It does satisfy the requirements for continuous variation to fit Darwin's views on blending inheritance and is a sort of solution. Pearson refined the equation to try to make it more practicable.[39] His modification probably represents an approximation to some principles of heredity, although it is not a particularly useful one. It is too general a statement and it is not based on experimental observations, but only on theoretical considerations. Even though the law is intuitively plausible, it is of little practical use because it fails to predict anything. No amount of clever mathematics as this could have led Galton or Pearson to deduce that inherited factors come in pairs (one from each parent) or that one of the factors is often dominant over the other in its expression. On the other hand, Mendel's formulation based on experimental results predicts that a particular character will appear (under certain defined conditions) in the next generation in one out of three siblings. This makes all the difference, because it points to a discoverable underlying mechanism giving rise to this ratio.

Pearson elaborated on Galton's equation and was very proud of his results. At the end of the article publishing Galton's modified equation he compared himself to Sir Isaac Newton by writing:

> *If Darwinian evolution by natural selection is combined*
> *with heredity then the single statement which embraces*
> *the whole field of heredity must prove as epoch- making to*
> *the biologist as the law of gravitation is to the astronomer.*

This is quoted again as a heading for Chapter 16. Hubris seemed to have made him overlook the fact that his formula was rather similar to a Taylor series[40], which has no theoretical significance but just postulates that the phenomenon in question varies continuously. The series can be used to express the relationship between two phenomena to any required degree of approximation.

Pearson was a great and original man, but could not bear to think of himself as being wrong in this case. It is so stupid to apply ideas rigidly when they do not fit the observations. We are often obliged in biology to start from rather hazy ideas; then clarification of the concepts may become an important part of the solution of the problem. However, Mendel's work was not a development of earlier ideas; instead, it was a new start. It marked a change from the qualitative observations of inheritance to the quantitative; from verbal descriptions to exact mathematical formulation. Pearson took offense and as an editor-in-chief of the journal that he and Galton helped to found called *Biometrika*, he would hardly allow any of Bateson's papers on Mendel's work to be published in it. Pearson was intolerant of any dissent from his own views. This is a serious fault for a scientist. For Pearson, inheritance was a matter of continuous graded changes based on the properties of gemmules that produced small variations for natural selection to work on; for Bateson it was a matter of discrete jumps that produced minor variations in body structures due to the inheritance of countable particulate units.

At the British Association meeting in 1904 in Cambridge there was a fierce showdown between the two schools of thought, between the Mendelian's versus the Biometrician's (called after Pearson's journal *Biometrika*). The meeting was held in the Sedgwick Museum that had just opened in 1903. The Pearson camp used the usual arguments against Bateson and the Mendelians: (1) There were often intermediate features found in many hybrid crosses that suggest blending inheritance as proposed by Darwin, and (2) there was the possibility that Mendelian results could be explained by other models than the one proposed by Mendel. Moreover, Bateson's methods were to be condemned as careless and his theories about underlying mechanisms as "cumbrous and indemonstrable." Bateson responded to this after the lunch break.

He flatly refuted the criticism and said *every science* that *deals with animals and plants will be teeming with discoveries made possible by Mendel's work.* He then likened Pearson's camp to "flat-earthers" who had described the paths of the heavenly bodies to harmonize with a theory of the flatness of the Earth, just as the Pearson group was harmonizing the facts of heredity with false ideas about blending inheritance.

Unlike political or religious battles, Galton was pleased to hear that no one was deliberately humiliated, no one was disgraced, no one was burned at the stake, and no one argued other than from the basis of their own experimental evidence, even if a few insults were hurled around. Pearson must have been partly persuaded by Bateson's arguments. He rose from the audience afterward to propose a 3-year truce to sort out these matters. Why would he do so if he had not believed that he was on the losing side? The chairman of the meeting, the Reverend T. R. Stebbing, a self-appointed "man of peace" after suggesting that compromise is a good thing went on: *You have all heard what Professor Pearson has suggested, but what I say is: let them fight it out.* This is what they went on to do.

The succeeding battle became much more intense than the earlier debate in physics about the nature of light. Is light made up of particles or waveforms? The two views are contradictory. There is one large group of phenomena that can be explained only on the basis of wave theory, and another large group that can be explained only on the particle theory. It appears that light can behave as both: when light arrives at, say, a photographic plate it behaves like a particle; when it is traveling between points it behaves more like a waveform with a definite wavelength. So, both views appear to be correct depending on the experimental circumstances. Scientists have to leave it at that, and wait for the future in the hope of attaining some wider vision that reconciles both views. Not so for heredity; there was to be no compromise. The Mendelians stated categorically that inheritance must be based on discrete countable particles; whereas the Pearson camp maintained that it was statistically based on the blending of gemmule-like properties. The controversy became quite bitter. Bateson was later to write to his wife Beatrice of one of his opponents, Walter R. Weldon: *If any man ever set himself to destroy another man's work that did he do to me....* Walter Weldon, of Weldon's dice fame[41], considered the basis of Darwin's hypotheses to be purely statistical and needed statistical methods for verification.

In general, when a scientific dispute rages for any length of time there is usually never a problem at bottom about mere words, but always a genuine problem about real objects and the names given to them—gemmules or genes.

It amused Galton that both sides claimed him as a founding father. Galton was claimed by the Pearson camp because he had published several papers and two books that promoted the idea of blending (or continuous variation) as a basis for inheritance. The Mendelian camp claimed Galton because of his studies on the Basset hounds. The Mendelians took Galton's results as support for a discontinuous type of variation as proposed by Mendel's original paper. However, in other respects, it was not so flattering to be proclaimed as a prophet by both camps; it was very unlikely that two distinct mechanisms had evolved to account for the basis of inheritance by sexual reproduction. It meant that his research work had been so ambiguous that both sides could read their own theories into his results. Professionally, Galton had well and truly sat on the fence.[42]

Gregor Mendel is quite rightly considered the founding father of modern genetics and his laws of inheritance have had a profound influence on the subsequent development of all biology. However, Galton's approach of using detailed family records to construct accurate and complete pedigrees, of studying affected sib pairs of both identical and nonidentical twins and devising refined statistical tests to assess the degree of inheritance between parents and offspring, have all withstood the test of time in the study of heredity. As an exact contemporary of Mendel (both born in 1822), it seems fair to conclude that Francis Galton started the field of quantitative genetics, that is, the genetics of continuous variation, except he was thinking incorrectly in theory with the idea of gemmules coming from all parts of the body and not of discrete particles (genes) coming from the sex organs. As for modern genetics, Galton is mainly remembered today for his statistical methods, the correlation coefficient being a good example. Mendel is the undisputed founding father of the theory and practice of the genetics of discrete variation. Mendel's ideas can also satisfactorily account for blending inheritance such as the intermediate skin color of children from a mixed marriage as being due to the interaction of three or more distinct genes.

The long reach of Mendel's gene

Mendel's discovery of the gene has led in time to the vast development of modern genetics, opening up such practical issues as medical therapeutics, gene therapy, personalized medicine, designer babies, and DNA fingerprints to provide unique identification of a person for legal or civil inquiries, improved agricultural crops, and improved industrial products.

When I was a medical student, the only way to treat diabetics with insulin was a messy business of extracting the hormone from the pancreas of cows or pigs, purifying it, and then injecting it into the diabetic patient. No human insulin was available at the time and some patients developed complications from the use of animal insulin. When the human insulin gene was isolated, it became possible to grow the gene in bacteria in large vats, which then synthesizes insulin on a commercial scale. This has led to the manufacture of human insulin for medical uses. Many other therapeutic agents are made similar to this, including growth hormone and some antibiotics.

Diseases such as cystic fibrosis or muscular dystrophy are due to faulty genes leading to the dream of replacing them—the so-called gene therapy. This process of genetic engineering is technically much more difficult than it sounds. It has worked for a defective gene that makes the child susceptible to multiple infections. This is an immunodeficiency disorder called severe combined immunodeficiency and is potentially fatal. It is a primary immunological deficiency in which there is combined absence of T lymphocyte and B lymphocyte function used for combating infections. There are at least 13 different genetic defects that can cause this. These defects lead to extreme susceptibility to very serious infections and the child has to live in a semisterile environment. This condition is generally considered to be the most serious of the primary immune deficiencies. Effective treatment by replacing the faulty gene in healthy stem cell transplants can cure the disorder. The future holds promise for gene therapy for several more types of these disorders. The child is freed from his restrictive environment and can roam the wide world. Even cleverer is to edit the faulty gene, or to modify the faulty region of the gene, and so restore normal function. This means the gene always remains in its correct position on the chromosome and under its normal controls; a difficult job to do is when trying to replace the whole gene from scratch. Progress is being made along these lines with one of the muscular dystrophies.

Therapy with some drugs can lead to severe adverse reactions even to the point of killing the patient. For example, a common group of drugs, the statins, are used to treat high blood cholesterol and heart disease. In rare cases, they can produce a severe muscle reaction (called rhabdomyolysis) in which the muscle breaks down releasing the cellular contents. This can kill the patient. There is a change in a separate gene that can be used to predict if the patient taking statins might go on to develop this severe muscle reaction. Alternative treatments can then be applied, opening up the field of personalized medicine where the best therapeutic agents can be selected to treat the patients by analyzing their genetic makeup. This is especially useful in some of the cancers.

The gene comes into the field of reproductive medicine to create designer babies. Instead of letting a mother give birth to a child with a horrible inherited disease such as cystic fibrosis or muscular dystrophy, the defective gene can be identified after *in vitro* fertilization when the embryo is just a ball of cells and then not placed back into the mother's uterus for growth. Instead, a healthy ball of cells selected by gene analysis can be implanted that has the chance to produce a healthy baby. This technique is now in use worldwide.

As a last medical example, the Government of the United Kingdom has embarked on the "100,000 Genome Project." The aim is to analyze the complete DNA sequence of 100,000 British subjects, concentrating on people with cancer and rarer metabolic disorders. It is hoped that this knowledge will unlock many more basic gene abnormalities that contribute to these disorders and develop more effective therapies to treat them.

Mendel discovered the gene by growing peas. Growing rice is another staple food crop especially for Asian populations. Poorer children in some of these regions run into vitamin deficiencies. Vitamin A deficiency is the leading cause of preventable blindness in children and increases their risk of disease and death from severe infections. Approximately, 250,000 to 500,000 malnourished children in the developing world go blind each year from deficiency of this vitamin, approximately half of whom die within a year of becoming blind. Golden rice is a genetically engineered rice plant that has the added gene for the manufacture of a precursor of vitamin A. It produces an edible rice grain with raised amounts of this precursor, thereby supplying the child with vitamin A. The Golden Rice project is currently being tested, in partnership with collaborating national research agencies, in the Philippines, Indonesia, and Bangladesh.

The results so far look promising. In 2013, Pope Francis gave his personal blessing to the Golden Rice project. This is important as he opposes the faction campaigning against the use of genetically modified crops. This faction is active in many countries where Catholicism is the main religion. One such country is the Philippines, where more than 80% of the population identifies as Roman Catholic, and field trials of Golden Rice are nearing completion. An official blessing of the church, therefore, could do a great deal to build support, allowing the Philippines to serve as a model for many of its neighbors on the potential health impacts of widespread availability and consumption of the golden grain.

As for scientific advances the use of genes would fill another book. More than 18 Nobel Prizes have been awarded to geneticists since 1907. Many of them are listed in the gene time line given at the end of this book. One advance pointing us in new directions involves computing. The chemistry of the gene has made us realize that it is a key informational molecule of life providing instructions on how cells should multiply, and to keep all the machinery of the cell in a working order. It has shown us a novel way of storing and manipulating information. Could it be used instead of electrons and silicon chips as the basis for a new type of computer? DNA might one day be integrated into a computer chip to create a so-called biochip that could push computers even faster. For certain specialized problems, DNA nanostring computers are faster and smaller than any other computer built so far. Furthermore, mathematical computations have been demonstrated to work on a DNA computer. Researchers at Caltech in California have created a circuit made from 130 unique DNA molecules, which is able to calculate the square root of numbers up to 15.

While still in their infancy, gene computers may be as capable of storing as much data as our personal computer. Genetic material for computers may be competing in the market place with silicon-based computers within the next decade.

chapter sixteen

Celebrating Mendel's discovery

If Darwinian evolution be natural selection combined with heredity then the single statement which embraces the whole field of heredity must prove as epoch-making to the biologist as the law of gravitation to the astronomer.

K. Pearson, mathematician and statistician (1857–1936)

Mendel discovered such a single statement that covers the whole field of heredity by sexual reproduction, and as endorsed by Karl Pearson, must be compared to Sir Isaac Newton, universally acknowledged as one of the world's greatest scientists. Mendel's heredity equations generalize from plants, to horses, to sea horses (a sort of marine fish), to sea eagles, and to humans. That a humble Catholic monk should be placed in the same scientific category as Sir Isaac Newton is bound to raise controversy. Many people attacked Mendel's results, including de Vries. It is shameful to have to report that one of the most virulent attacks came from a famous British geneticist Professor Sir Ronald A Fisher FRS (1890–1962) who published a 21 page article in 1936 listing his criticisms of Mendel's work:

> that the paper is only intelligible if the experiments reported in it were fictitious; that the data of the later experiments were biased strongly in agreement with expectation; and that the data of most, if not all of the experiments have been falsified so as to agree closely with Mendel's expectations.

This is the most damning criticism one can make of a scientist's work and ruins their reputation ever after. It is the one thing that the scientific community will not forgive, deliberately fabricating results. Mud sticks and many people even now react to Mendel's name as the one who faked his data. Fisher wrote Mendel's results were too good to be true. In large number of plants Mendel studied there would be sampling errors and Fisher could find no signs of them; therefore, the results must have been falsified. Fisher's analysis of Mendel's results was necessarily incomplete because all Mendel's notebooks and data files were destroyed when Mendel died to make room for the incoming abbot of the monastery. So, all Fisher had to work on was the paper published in the Brünn Scientific Society of 1866, which was a summary of the previous 8 years work. Mendel wrote that he discarded unhealthy plants because of poor seeds or pollen. Assault on Mendel's integrity was all the more surprising because Fisher did not attack any of the later scientists who found even closer ratios of 3:1 than Mendel (Tschermak in 1900, Garrod in 1902, and so on). Mendel died in 1884, so there was no possibility of defending himself.[43] Envy and egocentricity are some of the besetting sins of scientists. Mendel may be more akin to the poets such as William Blake (1757–1827) who wrote in his long poem *Jerusalem* that *I will not reason and compare, my business is to create*.

Later (2008), one of Fisher's students with four colleagues set up a sort of Kangaroo Court to judge Mendel's work. They published a book[43] sifting the evidence and all concluded that "Mendel was not guilty of fraud." No doubt Mendel's spirit could now rest in peace.

Mendel may have well pondered on this famous scene described by Darwin in the last chapter of *The Origin of Species*:

> *It is interesting to contemplate a tangled bank, clothed with many plants of many kinds, with birds singing on the bushes, with various insects flitting about, and worms crawling through the damp earth....have all been produced by laws acting around us. These laws...being Growth with Reproduction; Inheritance which is implied by reproduction....*

Mendel did more than contemplate this tangled scene. Out of its complexity by a mixed process of intuition, observation, and analysis, he abstracted the key elements that explain some of the basic aspects of reproductive inheritance. Mendel disentangled some of the principles of heredity and gave us a glimpse into a new aspect of the truth that would open up whole new fields of endeavor, such as designer babies, genetically modified plants, and molecular drug discovery. He succeeded in extracting the general ratios of inheritance by his skillful use of experiments and measurements. There is no logical path to these principles; it can be done by a sort of creative intuition, involving irrational elements based on a deep sympathetic understanding of the things under study.

Or was he just lucky, as some scientific discoveries are? Most people would think that he must have had some specific ideas already in mind to sustain him for 8 years of solitary labor. Eight years of producing and describing hybrid pea plants does not sound a very exciting way of spending one's time, but such careful measurements belong to the scientist's work, just as the hammer and chisel belong to the work of the sculptor. Michelangelo's patient and repetitive chipping away at blocks of marble produced some of the most inspiring art objects that we possess. The test of a vocation is the love of the drudgery that it involves.

After 8 years work and analysis of about 10,000 pea plants, Mendel had the following insights that best accounted for his observations. The appearance of two different characteristics (tall or dwarf plants; or variation in flower color or position on stem etc.) in the second generation (F2)[44] after cross-breeding plants of the first generation (F1) occur in the ratio of 3:1. This has been called Mendel's first law or principle of inheritance. He actually called it these himself, but they are more like postulates. His insights were that:

1. Each adult pea plant has two factors of heredity for each trait he was studying; one comes from the male parent in the pollen, the other from the female parent in the ovule (or egg cell). Mendel called each hereditary unit an Elemente or Zellelemente that can be translated in English as a unit factor or element.
2. During the formation of the pollen grain and egg cell (or ovule) by the parent plants each factor of the adult pair separates out and one of the factors is transmitted randomly and independently into each of the reproductive cells (pollen or ovule) used for fertilization to create the next generation of plants.
3. Consequently, each pollen or egg cell carries only one of Mendel's factors for each trait.
4. The union of pollen and egg cell restores at random the full units of heredity (back to two factors, one from each parent) for each of the plant's inherited feature and goes to the formation of the first cell of the new plant (or animal if considering sperm and egg).

(The inheritance of sex is also random, usually occurring in a ratio of 1:1 based on difference of a single factor, the Y chromosome.)

Most people would now argue in a verbal language, but from then on Mendel's crowning glory was to see that it could be formulated in algebra. The beauty of doing this is that one can employ exact thoughts with logical rigor and one has to state one's initial assumptions clearly; there is no room for woolly thinking as there is in a verbal language. Simply by using the fundamental laws of algebra (the laws of addition, multiplication, and distribution), Mendel built up his model for inheritance. Many people were dazed by the details of Mendel's algebra, and even for Galton, it was a rather obscure branch of mathematics, seeming to be a mixture of the binomial theorem with analysis of combinations and permutations.

Mendel first defined his terms as (A) representing, for example, a hereditary element coding for a tall plant and the symbol (a) to represent an element coding for a dwarf plant.

Then his hypothesis assumes that each plant possesses two factors determining the height of the plant: one coming from the egg cell, the other coming from the pollen that fertilized the egg; these factors we have called (A) and (a).

At the start Mendel was using pure inbred lines of plants that he demonstrated to be pure by self-fertilization (which the edible pea does readily due to the shape of its flower). On self-fertilization the tall plant always produced a tall plant; and the dwarf plants always produced short plants. The initial tall pea plant was therefore assumed to carry (A + A) and the initial dwarf plant to carry (a + a).

To be precise, Mendel was not quite sure whether this was, in fact, an additive relationship, so in his original paper he represented it as A/A or a/a. Then when he crossed the tall with the dwarf plant to produce the first generation of hybrids (F1) his hypothesis assumed that there would be independent distribution of the two elements from each of the parent plants into their pollen and egg cells; so for this first generation he could only obtain four combinations of elements for four types of plants all carrying the same hereditary units (A + a) as

$$(A + a), \text{ or } (A + a), \text{ or } (a + A), \text{ or } (a + A)$$

These all turned out to be tall plants; the dwarf trait (a) seemed to have disappeared.

Now crossing (A + a) to (A + a) leads to the second generation of plants (F2) when the dwarf trait reappears. It had not been swamped out by the tall trait (A).

Mendel's first law states in algebra:

Hybridizing the first generation (F1) would lead to the second generation (F2) of inherited factors as

$$\text{F1 hybrids: } (A + a) \times (A + a) \Rightarrow \text{F2 hybrids: } AA + 2Aa + aa \qquad (16.1)$$

This resembles a partial solution of a quadratic equation learned in school days. It is a shame that mathematics teachers at schools do not point out the connection between quadratic equations and this theory of inheritance of Mendel relating to an equation describing real-life events.

What it says in words is: there is random distribution of hereditary elements into the gametes (egg and pollen) and then random combination of gametes in binomial proportions to form the fertilized offspring (sometimes called a zygote) with individual plants carrying one of (A + A), two of (A + a), and one of (a + a), so therefore in the ratios of 1:2:1.

These are not obvious relationships, so Mendel deserves all the more admiration and credit for discovering them.

This is not an abstract model for his experimental results but turned out to be an actual representation of what happens; the (A) and (a) elements, unlike gemmules, actually exist, which we call genes nowadays, and they combine according to the aforementioned algebraic expressions.

Where did the Equation 16.1 come from? It represents the hereditary makeup of the four possible offspring of the second generation. It looks similar to an ordinary quadratic equation (a binomial expansion to the power of 2), which to remind the reader looks like this:

$$(A+a)\times(A+a) = AA + 2Aa + aa = A^2 + 2Aa + a^2 \qquad (16.2)$$

A and a are the two terms of the binomial equation and the multiplication sign indicates cross-fertilization of two pea plants.

The last equation (16.2) has been appropriated by population geneticists and is used today as the Hardy–Weinberg equation[45] when considering allele/genotype probabilities that are multiplicative for combined events.

Mendel's equation—the middle equation (16.2) is not straightforward algebra; AA does not signify a multiplication of $(A \times A = A^2)$. Mendel means $(A + A)$ to signify the parental source as egg or pollen. So, Mendel's expression (16.1) cannot be simplified any further to the binomial form of

$$A^2 + 2Aa + a^2 \text{ factorizing to } (A+a)\times(A+a)$$

Surprisingly, Mendel then wrote

$$A/A + 2Aa + a/a \Rightarrow A + 2Aa + a$$

How did he simplify it to this? He is perhaps switching from genotype (*Elemente*) numbers to phenotype numbers (*Merkmale*) in one line.

Where does the ratio of 3:1 that Mendel discovered come from? Another great insight of Mendel was that element (A) is dominant over (a), which he called recessive (alleles in modern genetics). This means that if (A) codes for a tall plant and (a) codes for a dwarf plant and if the two factors are inherited together as (A + a) then the plant will be tall because (A) overrides the effect of (a); or in other words (a) is recessive to (A). If (a + a) occur together, that is two recessives, the plant will be dwarf. Why did Mendel propose this unlikely idea of dominant and recessive inheritance? He needed these terms to make more sense of his experimental observations for his second generation of hybrids. Mendel was correct and we now know that a dominant element can make enough of its product (RNA) not to need the recessive element for the tall trait to appear.

To return to the hybrid crosses, if a plant inherits (A + a) it will be tall, if (a + a) it will be dwarf. Therefore, adding up the terms of Equation 16.1 above we get

aa—one dwarf plant
AA—one tall plant
2Aa—two tall plants

Giving a ratio of 3 tall plants to 1 short, that is, a ratio of 3:1.

And, EUREKA, this is what he found experimentally. He found a new biological ratio that was found later to extend from plants to fish, flesh, fowl, and to flies. This binomial type of algebra seems to enter genetics effortlessly and it is amazing that the equations tie in so well with the data. It is interesting to speculate, which came first in Mendel's mind: an intuition that a binomial-type equation is going to work for this system; or the experimental data gathered over 8 years was subsequently found to fit a binomial equation. I personally believe his equation came first as a creative insight and then the work came after to verify his theory. The experimental 3:1 ratio would therefore count as a prediction if his theory were correct.

The language of heredity, similar to physics and chemistry, can be written in mathematics (binomial and probability theory). Moreover, if a gene is defined as a unit of heredity that controls a particular inherited characteristic of an organism, such as a tall or dwarf plant, then we must acknowledge that Mendel discovered the "gene," which in his terminology he called *Elemente*. Much later in 1909, the word gene was introduced by Wilhelm Johannesen (1857–1927), a Danish professor of botany and plant physiology, which had some of the properties of *Elemente*. Mendel then went on to formulate his second law of inheritance involving the coinheritance of separate plant characteristics called the law of independent assortment of inherited characteristics. These are seen in dihybrid crosses.

Let letters (Aa) and (Bb) now represent two hereditary features of the plant determining such things as tall (A) and dwarf (a) as before, plus smooth (B) or wrinkled pea coats (b). The algebra then became more complicated as he went on with the third and fourth crosses using different dihybrids, producing horrible equations similar to this (please only glance at it):

Mendel's second law of inheritance: the expected inheritance of two independent traits is

$$ABAB + ABAb + ABaB + Abab + AbAB + abab$$

No wonder the fainthearted could not follow him; the nonmathematician is often seized by a mysterious shuddering when he sees such expressions as this by a feeling not unlike standing on the edge of a high precipice (Was this one of the reasons why Darwin ignored the work?). Even for some fairly experienced algebraists it is not easy to understand the full implications of these expressions. What it says in words is: the different features in the plant represented by genotypes Aa and Bb and so on are coinherited independently of each other (provided there is no chromosomal linkage).

The originality of using algebra is that it can show connections between things that are not obvious at first sight. A difficulty is that algebra changes the relative importance of the terms found in ordinary language. It still is essentially a written language with its own dictionary and syntax, and after defining its terms the consequences are not negotiable except by logic. It endeavors to exemplify in its symbolic structure the pattern of natural occurrences that it is intending to describe for the reader. It is astonishing to see how important for the development of Mendel's hypothesis a modest-looking symbol like (A) and (a) may become. The presentation of such a simple idea can by elaboration lead to a complex train of abstract ideas that follow from it eventually leading to one of the fundamental laws of nature that affects us all (i.e., the chances of our next child suffering from cystic fibrosis or sickle cell anemia). Most of the other genetic human diseases are more complicated than this involving change in many genes, but the so-called Mendelian disorders are generally inherited as a change in a single gene.

It is remarkable that the tools of simple algebra such as this can be portrayed in a pattern of nature. Newton also used simple mathematics for his law of universal gravitation—one might expect the laws of nature needed to be written in a more complex form. Mendel was using his symbolism almost as a pictorial representation of the relations of plant characteristics to each other during inheritance. Mendel's hypothesis made beautiful sense of all the observations. It was quite extraordinary to see how the results of his experiments agreed almost exactly with the calculations deduced from his model. Mendel had done what all dream of doing in biology to provide sufficiently rigorous and precise descriptions of important relationships that could be expressed in mathematical language and could predict the future events. All this was done for the complex subject of heredity. He was the first person to show that it was possible to formulate general biological laws by mathematics; in the same way that Newton was the first to frame the law of universal gravitation in mathematics. Reliable results in science are best secured by unambiguous statements. This is where we cannot do without the precision and clarity of an abstract mathematical language. Using only two general ideas, that of the algebraic variable and that of algebraic form (the equation), Mendel described a fundamental aspect of nature.

His formulation initially represented a proposal, deduced from a far-from-simple mathematical analysis of the experimental observations of his plant studies. With more experimental findings using different systems, would his equations still hold up? It is a paradox that all exact science is led by the idea of approximation. Scientific experiments are never finished; their resulting discoveries always need future modification. So, how far would Mendel's algebraic description of the laws of heredity take us before a revision is needed? Newton's laws of motion needed revision about 230 years after their publication by Einstein.

The problem with theories, however imaginative they may be, is that they take us beyond direct measurements in a way that cannot be foretold *a priori*. The danger comes at a time when the experimental results no longer agree with the theory and one fails to revise the latter. There have always been a number of erroneous theories on which a vast quantity of labor has been wasted; yet many problems that were first rejected as meaningless by the keenest critics were eventually seen to possess the greatest significance. A hundred years ago, physicists considered it meaningless to ask for the mass of a single atom—an illusory problem not open to any form of scientific measurement. Today, the mass of an atom can be measured to within its ten-thousandth part.

Many beautiful theories have been demolished under more stringent testing. For the scientist replication of experiments and consistency with other facts is synonymous with accuracy. Mendel went on to test his theory by crossing the third and fourth generations of hybrid plants, by crossing plant characters such as tall plants and smooth peas crossed with dwarf plants and wrinkled peas, and so on. He studied up to ten generations of hybrids. He then predicted what the results should be according to his model in these further generations. His predictions matched quite closely the actually observed numbers of inherited features in the further generations of plants. His laws (or principles) of heredity (independent segregation of genes into gametes and independent assortment of gametes into zygotes) is the holy grail of the biologist. This is to identify those universal laws, which connect some of the smallest entities of life, the genes or molecules, to features of the whole animal or plant and then to the societies they form. New biological phenomena can be predicted from these laws by logic, and the observed effects can be explained by them at each level of complexity. There is no logical path in pursuit of these laws; they may arise by intuition framing the right concepts, based on a sympathetic and correct interpretation of the experimental observations.

Mendel compares to Newton?

Think of the reach of Newton's three laws of motion and his law of universal gravitation. His three laws of motion are stated in his great book *The Mathematical Principles of Natural Philosophy* (1725) as follows: (1) Every body continues in its state of rest, or of uniform motion in a right line unless it is compelled to change that state by forces impressed on it, (2) change of motion is proportional to the motive force impressed on it, and (3) to every action there is always opposed an equal reaction.

Newton's laws can be easily visualized as acting on a billiard table. A billiard ball remains at rest unless struck by the cue and then it will travel in a straight line (Law 1); by the ball's force it can cause another ball to change its motion (Law 2); and the action and reaction of the two balls that collide are equal and opposite (Law 3). These laws follow on from Galileo's early experiments on moving bodies and Law 1 is similar to Galileo's formulation. They have had extensive use in all types of mechanics, including predicting the flight paths of aircraft and sending rockets to the Moon and Mars. In view of the title of Newton's great book, it is curious that he used prose and not mathematics to formulate them; whereas Mendel was able to formulate his laws in algebra. Newton's calculus (that he invented contemporaneously with Leibniz, but claimed priority) for the study of moving bodies would have been ideal to use. Laws 1 and 3 can easily be expressed in calculus and just for fun might read like this:

Law 1: If $\sum F = 0$, then $dv/dt = 0$; or in words, if the sum of vectorial forces acting on a body is zero then the change in velocity of that body will be zero.

Law 3: $dp1/dt = -dp2/dt$; if a body p1 collides with another body p2 there will be an equal and opposite force acting on p1.

It is much easier to work out the implications of his laws when they are written in calculus than when written in words.

Newton did use mathematics, but not the calculus, to formulate his law of universal gravitation, which states that two objects (an apple and the Earth, for example, or the Moon and the Earth) attract each other with a force directly proportional to the product of their masses divided by the square of their distance apart. This is the inverse square law. In mathematics, it reads as

$$F = \frac{G \cdot m1 \cdot m2}{d^2}$$

where:
   F is the force of attraction between mass m1 and m2
   d is the distance between them
   G is a gravitational constant

This has been hailed as one of the greatest of all scientific generalizations. Its mathematical deduction (involving no experimental work similar to Galileo or Kepler had done) has allowed immense tracts of experience to be gathered together under a unified theory; from the behavior of falling bodies on Earth, to the Moon's motion about the Earth, to the revolutions of the planets about the Sun, to the movements of the tides and to the trajectory of comets; and all these phenomena can be mathematically described with great accuracy from the basic postulates of Newtonian mechanics and gravitation. So, it is no longer possible to doubt that they are correct within certain limits.

However, the fundamental basis of gravitation still remains a mystery. Even today we do not fully understand the nature of the gravitational force; some modern ideas postulate an elastic curvature in space distorting the geometry of space-time, that is, the billiard table surface is curved, so balls will roll of their own accord to the lowest point.

Newton had to postulate it as a hypothetical force that can act through empty space—thus acting at a distance. It is as though a stationary billiard ball could be set in motion with no contact with another moving ball. Newton needed a God to impose such a force acting through empty space. He had no idea of the nature of it or how it operated; he had to resort to vague terms such as an "elastic spirit" that pervades and lies hidden in all of the physical matter.

With regard to originality, Mendel formulated his laws of heredity *de novo*. Newton admitted that he could see further than others only by standing on the shoulders of those who had gone before him (mainly using data of Galileo and Kepler[46]); whereas Mendel had no shoulders to stand on but his own. Of course, many botanists before Mendel had studied problems of plant hybridization, although none of them had thought about the process in an algebraic way Mendel did. His was an entirely new theory; and he managed to express phenomena in mathematics representing the observed regularities in the pattern of inheritance in plant hybrids. He postulated an underlying mechanism of dominant and recessive traits that linked the observed facts with his theoretical equations, which turned out to be both correct and experimentally useful. Using Mendel's laws we can predict the average occurrence of many uncommon diseases in families where they segregate. Despite being a priest he did not need to invoke a God for his laws of heredity, whereas Newton the scientist needed a God to explain his law of universal gravitation.

In many ways, one can estimate the originality of a scientist's work by the public reception it receives. Thanks to Lamarck and others the theory of evolution was very much "in the air" when Darwin published *The Origin of Species*. It quickly sold out and a second edition was published in January 1860 to meet public demand; the third edition came out in 1861, with three more editions during Darwin's lifetime. In a similar manner, Newton's laws of motion and universal gravitation were "in the air," thanks to the *Discorsi* of Galileo in 1638. Newton quickly reaped the honors for his work being elected Lucasian Professor of Mathematics at

Cambridge at the age of 26, a post he held for the next 32 years; elected as the President of the Royal Society in 1703; knighted by Queen Anne in 1705; and when he died in 1727 it was an occasion for pomp and circumstance with a ceremonial burial in Westminster Abbey under an elaborately carved marble tombstone commemorating his discoveries, the issuance of gold and silver medals, more life-size statues in marble, and many laudatory poems.

Mendel's work was definitely "not in the air" at the time he published it. Perhaps as many as 35–40 top scientists in Europe saw or read the paper and then promptly ignored it. If a brand new observation is placed in front of one, it can be very difficult to assess its importance if there is no previous background information to go on. For Mendel, there was nothing in the scientific literature to precede his theory; there was no background work to direct people's thoughts in a particular direction and one needed statistics and the algebra of the binomial theorem to fully understand his ideas. It lay fallow for about 35 years before the rest of the field caught up with him in the early 1900s.

Comparisons are odorous as Dogberry says in Much Ado About Nothing, but I think Mendel's work was more original than Newton's, although Newton's work has had more practical applications and spans the whole reach of the universe (and, of course, includes his work on optics). Mendel's work is closer to humanity, his work leading to all the scientific developments listed in the gene time line at the end of the book. This includes genetic engineering, DNA as the information molecule of life, the double helix, and the practical spin-offs that are described in the Long Reach of the Gene.

Perhaps we do not fully appreciate the dedication and labors of our pioneers that have led to all the developments that we take so much for granted. From Mendel's botanical studies, we can gain inspiration from his work by thinking of a day when we started to do something slightly unusual.

Epilogue one: Why Darwin might disregard Mendel

Faith is a fine invention
For Gentlemen who see!
But Microscopes are prudent
In an Emergency.

Emily Dickinson (1830–1886)

After publication of *The Origin of Species* in 1859, Bishop Samuel Wilberforce of Oxford was the first of a long line of distinguished clerics to attack Darwin publicly. In the famous Oxford Debate of 1860, the Bishop stated that *Darwin's endeavours to prop up his utterly rotten fabric of guesses and speculation by a fanciful tissue of lies* and *whose mode of dealing with nature is reprobated as utterly dishonorable to the Natural Sciences*; went on: *I should like to ask Professor Huxley, who is sitting over there, as to his belief in being descended from an ape. Is it on his grandfather's or his grandmother's side that the ape ancestry comes in?* Later, Huxley avenged this slur on his parentage when in 1873 Bishop Wilberforce unfortunately fell off his horse onto his head and was killed. Huxley's pithy response was "Poor dear Sammy! His end has been all too tragic for his life. For once reality and his brains came into contact and the result was fatal."

Cardinal Manning (1808–1892) joined the fray against evolution and declared his abhorrence of the new view of nature describing it *as a brutal philosophy—to wit there is no God and the ape is our Adam*. He violently opposed Darwin's view that man had descended from *a hairy quadruped, furnished with a tail and pointed ears, and probably arboreal in its habits*. Dean Burgon (1813–1888) of Chichester preached a sermon to the University of Oxford, warning the students that *those who refuse to accept the history of creation of our first parents and are for substituting the modern dream of evolution in its place, cause the entire scheme of man's salvation to collapse*. The Reverend Father Bayma, a professor at the Jesuit College of Stonyhurst, wrote in the *Catholic World* that *Mr. Darwin is, we have reason to believe, a mouthpiece or chief trumpeter of that infidel clique whose well known object is to do away with all idea of God.* And Cardinal Wiseman (1802–1865) having read *The Origin of Species* declared that:

> *It is one of the most detestable theories I have ever come across. The purely mechanistic explanation as to our origins is a personal affront. It repudiates final causes and all he seems to be saying is that man is no more than a transmuted ape.*

European theologians were equally vociferous, and Monseigneur Segur (1820–1881) referring to Darwin and his followers as impersonators of God's will went into a form of mild hysterics:

> *These infamous doctrines have for their support the most abject passions. Their father is pride, their mother impurity, their offspring revolutions. They come from Hell, and return there, taking with them the gross creatures who blush not to proclaim and accept them.*

Eventually, Pope Pius IX (1792–1878) felt impelled to join the attack and thanked a French Catholic physician (Dr. Constantin James) for his book that *refutes so well the aberrations of Darwinism,* His Holiness adding:

> *which is so repugnant at once to history, to the tradition of all people, to exact science, to observed fact, and even to Reason herself, and would seem to need no refutation, did not alienation from God and the leaning towards materialism eagerly seek to support everyone in this tissue of lies...*

Not all the criticism was in the form of irrational abuse. One eminent professor (probably Professor Richard Owen) wrote that:

> *The facts seem to tell me that animal species have been constant for thousands and thousands of years, and will never change so long as conditions remain constant. Change the conditions and I agree old species may disappear and new species may appear in their place and flourish. But how; and by what mechanism? I say by creation. This is an operation quite beyond the powers of a pigeon fancier or animal breeder like Mr. Darwin. There is a mystery here about a force that I cannot*

> *imitate or comprehend—but there is a design and pur-*
> *pose working in the world which I can attribute to a*
> *power which by one name I can call God.*

There was so much adverse criticism of Darwin's work that might render him quite prejudiced against any scientific studies done by a Catholic priest.

Epilogue two: Scientists quarrel—Romanes and Wallace

Answer not a fool according to his folly, lest you become like him.

Proverbs 26:4

The quarrel between Darwin and Galton is described in Chapter 6. Here is another.

Romanes decided to write a book about the ideas of his great mentor Darwin. He thought it would be nice to have a photograph of Alfred Wallace, the codiscoverer of natural selection, placed somewhere near the front. He accordingly wrote to Wallace to ask for a suitable one and was quite offended to have his request turned down sharply.

It is true that years before (in 1886) they had quarreled in public about an obscure point of how natural selection might operate by "physiological selection." Tempers were eventually lost and it ended with Romanes ridiculing Wallace in public. He wrote that *we encounter the Wallace of spiritualism and astrology, the Wallace of incapacity and absurdity.* Wallace made no reply, but it had touched him on two sore spots. Wallace was not as well educated as Romanes, not having a university degree and coming from a relatively impoverished family. Academically, he rose to be a master at a collegiate school in Leicester. From the start, Wallace was keenly involved in spiritualism. In time he became an ardent believer in the occult and spiritual world and all the different types of contacts that were made with us ordinary mortals from the other world. There was an element of danger in all this because "calling up the dead" was considered a blasphemous act and was banned by the parliament in the Witchcraft Act of 1735. The last spiritualist to be imprisoned for contravening this act was Helen Duncan in 1944. In 1941, she called up the spirit of a drowned sailor from HMS Barham. The ship had indeed been sunk in 1941, but this was meant to be kept as a state secret. Was she, in fact, a spy for the Germans?

There was an enormous public outcry against the case and the Act was repealed by Parliament in 1954.

Wallace started to organize séances in his own home and attended other people's houses; he heard tables rapping, saw chairs floating off into space, read letters sent from the other world, and encountered strange materializations. At one séance he was invited to go under the table while the medium (a Mr. Home) played a sweet tune on an accordion with one hand, the other hand resting on the table. The room was well lit and Wallace distinctly recorded that, when Mr. Home removed his hand from the instrument it went on playing and a disembodied hand appeared to hold the instrument, while Mr. Home's both hands were now on the table. Galton saw this séance too and was mystified.

At another séance, Wallace saw a materialization of a white-robed female figure that would wander vaguely around the room and allow her face, hair, and ears to be examined; while the medium dressed in black (a Mrs. Cook) was all the while sunk in a deep trance. Sometimes the materialization would take the form of a tall stately East Indian figure in white robes, a rich waist band, sandals, and a large turban, snowy white and disposed with perfect elegance. He stalked grandly around the room in a detached manner and bowed gracefully before several of the guests.

Put like this the séances sound rather ridiculous, but there was a serious intention behind it all. With the rise of agnosticism in the nineteenth century as science and technology became more dominant in society, many intellectuals, including Wallace, Darwin, and Galton rejected Christianity. However, some felt the need to have something in its place. Wallace in particular required a superior intelligence for his idea of evolution that guides ever more perfect adaptations of animals and plants to their natural environment. There had, in Wallace's view, to be some supernatural force driving this and its scientific investigation seemed to be an important and worthwhile aim. Wallace was not alone in this and spiritualism became a popular movement in the mid-nineteenth century. A Society for Psychical Research was established at Cambridge University in Isaac Newton's old college of Trinity; and many first class scientists were recruited to its ranks. Their aim was to record, measure, test, and experiment on the various manifestations and materializations that emanate from the other world. This activity gained a wide following among many different classes of Victorian society and there were plenty of opportunities to engage in spiritualistic experiences at séances in the London houses of the fashionable and wealthy.

Wallace only attended séances where the mediums were unpaid and that were performed under strictly controlled conditions. He was soon so much convinced by it that he started to write books and pamphlets on the subject. *Miracles and Modern Spiritualism, The Scientific Aspects of the Supernatural,* and *A Defense of Modern Spiritualism* all attested to his firm belief in the occurrence of these other-worldly experiences.

Some of his friends were aghast on reading this. Huxley wrote quite bluntly to him:

> *I am not disposed to issue a Commission of Lunacy against you. It may be all true for anything I know to the contrary, but I really cannot get up any interest in the subject. I never cared for gossip in my life, and disembodied gossip, such as these worthy ghosts supply their friends with, is not more interesting to me than any other.*

Another friend, Professor John Tyndall (the physicist), was equally blunt:

> *I see the usual keen powers of your mind displayed in the treatment of this question. But mental powers may show itself whether its material be facts or fictions. It is not the lack of logic I see in your book but a willingness that I deplore to accept data which are unworthy of your attention.*

Charles Darwin had also attended séances on the invitation of his brother Erasmus in his London house. Tables dutifully rapped, chairs jumped about, and fierce fiery lights darted about the darkened room. It was all so hot and tiring that Darwin left early. He kept a fairly open mind on the subject, although could not believe it was anything more than mere trickery; *Lord have mercy on us all if we have to believe in such rubbish* is what he said.

Galton in his pursuit of the truth was also quite interested at the start and attended several séances, including the one with the disembodied hand playing the accordion under the table. He thought it was absurd on the one hand but incredibly clever on the other to arrange all the stage effects so convincingly. He started to investigate the matter scientifically during the performances. He wanted firm evidence that the manifestations occurred independently from the medium, even if the latter had to be locked up in a wire cage and the lighting and conditions of the room were to be altered. One imposter (a certain Mrs. Hayden) was uncovered in this way. In fact, she readily admitted to the fraud when interrogated under pressure. However, Galton became too experimental in methods and he deeply offended the spiritualist fraternity who gave him up as an apostate. He was no longer invited to any further séances and he quickly lost interest in the subject.

And what of Romanes who had taunted Wallace for being a spiritualist? He was double-faced. In 1879, he wrote several letters to Charles Darwin describing his experiences of spiritual phenomena. At one séance he saw hand bells moving about the room; a human head and face wafted above a table with the face having mobile features and eyes.

Another bell placed on a piano some distance away was taken up by a disembodied hand and rung, and then carried about the room while the piano sounded. Romanes wrote in another letter about his conviction of the truth of these facts and of the existence of a spiritual intelligence, of a mind without a brain. These experiences had completely altered his conception of the world.

Romanes was clearly a believer, but rather ashamed to confess it to his fellow naturalists some of whom he knew would pour ridicule on him. In the same way he did not want it to be known generally that he wrote poetry because it might lower the regard his scientific colleagues might have for him.

In Wallace's view Romanes was a pure hypocrite because he was not ashamed of making use of the ignorant prejudice against spiritualism as a way to attack Wallace in a scientific argument. Whereas Wallace at least had the courage of his convictions to declare publicly his beliefs in spiritualism.

That could be one of the reasons why Romanes never got the picture to put in his forthcoming book on Darwinian theory.

Epilogue three: Mendel in his own words[*]

Introductory remarks

Artificial fertilizations, undertaken in ornamental plants for the purpose of creating new color variants, were the motivation for the experiments to be discussed herein. The remarkable regularity with which the same hybrid forms always recur as often as fertilization involved the same species, stimulated further experiments designed to follow the progeny of the hybrids.

To this end, careful observers such as Kölreuter, Gärtner, Herbert, Lecog, and Wichura have devoted a portion of their lives with indefatigable perseverance. Notably, Gärtner in his work *The Production of Hybrids in the Plant Kingdom* has published very valuable observations, and in recent times Wichura has published thorough investigations on the hybrids of willows. Nobody should be surprised that as yet it has not been possible to establish a generally valid law on the formation and development of hybrids; and who knows how to evaluate the extent of the task and to appreciate the difficulties inherent in experiments of this type. A final decision can follow only when detailed experiments are documented from the most diverse plant families. Whosoever surveying the publications on this subject will reach the conviction that among the numerous experiments, none were conducted to such an extent and in such a manner to make it possible:

- To determine the number of the different forms occurring among the progeny of the hybrids.
- To order these forms with certainty in the individual generations.
- To determine their numerical relationships.

[*] Excerpts from a literal translation of Gregor Mendel's Experiments in Plant Hybridization. From the Transactions of the Natural History Association of Brünn Vol 1V (1866), translator: Professor J M Opitz, University of Utah, 2017.

However, some courage is required to undertake such an extensive under-taking; yet, this seems to be the only correct way to attain, finally, the solution of a problem, which is of inestimable importance for the develop-mental history of organic forms.

The present treatise considers the execution of such a detailed experi-ment. This was confined to a smaller group of plants and now, after 8 years, is essentially concluded.

Selection of the experimental plants

Value and validity of each experiment are determined by the suitabil-ity of the materials employed for that purpose and the effective applica-tion of the same. In the present case as well it cannot be indifferent of which plants are selected as subjects for these experiments and how these experiments are conducted.

The selection of the plant group to serve for experiments of this kind has to occur with the utmost care, so as to avoid from the beginning all results as questionable.

It is necessary for the experimental plants

1. To manifest reliably differentiating characters.
2. During flowering the hybrids of the same have to be protected or easily protectable from the influence of all foreign pollen.
3. The hybrids and their progeny in successive generations should not suffer notable impairment of fertility.

Unrecognized adulteration by extraneous pollen, which may occur during the course of the experiments, may lead to completely errone-ous conclusions. Diminished fertility or total sterility of some forms, which occurs in the progeny of many hybrids, would severely impair or invalidate the experiments. In order to recognize the relationships of the hybrids to each other and to their progenitors, it seems necessary that the members of each resulting category in each generation be exhaustively subjected to analysis.

From the beginning special attention was paid to the *leguminosae* because of the characteristic structure of their blossoms. Experiments conducted with several types of this family led to the result that the genus *Pisum* complied sufficiently with the necessary requirements. Several completely distinct forms of the genus possess constant char-acteristics easily and securely classifiable and yield completely fertile progeny from their reciprocal crosses. Also, it is not easy for extrane-ous pollen to interfere because the generative organs are closely packed inside the keel, and the anthers burst already in the bud, so that the stigma is covered with pollen even before the opening of the flower.

This circumstance is of particular importance. Further advantages of this plant are the ease of its propagation in the open and in pots and its relatively short life span. Artificial fertilization is somewhat involved but almost always successful. For this purpose, the not yet fully developed bud is opened, the keel is removed, and each stamen is taken out carefully with tweezers, so that the stigma can then be covered immediately with the foreign pollen.

From several seed companies a total of 34 more or less distinct kinds of peas were obtained and subjected to a 2-year study. In one batch, among a larger number of identical seeds, several ones of significantly deviant form were noted. However, these did not vary in the next year and were completely identical to another type obtained from the same seed company; without doubt these seed had been mixed accidentally. All other sorts uniformly yielded the same and constant progeny; in any event, in both preliminary experimental years no substantial deviation was noted. For the purpose of fertilization, 22 of these were selected and planted annually during the entire initial exploratory period. They all stood the test without any exception.

Schedule and order of the experiments

The several pea types selected for study differ in the height of the plant, in the size and shape of the leaves, in the position, color and size of the flowers, and in the pigmentation of the seed coat and of the albumen. However, a part of these traits does not allow a clear cut separation because the distinction may be difficult-to-determine as a *more or less* nature. Such traits were not useful for the individual experiments, which had to be confined to clearly distinct and contrasting plant characteristics. It was the results then which had to show whether in a hybrid union all of them showed the same behavior to allow evaluation of those traits of a regular but subordinate nature.

The characters selected for the experiments were the following:

1. The *difference in form of the ripe seed*: These are either spherical (or round) with surface indentations, if present, always shallow; or, they are irregularly edgy, deeply wrinkled (*P. quadratum*).
2. The *difference in color of the albumen (endosperm) of the seed*: The albumen of the ripe seed is either pale yellow, bright yellow, or orange; or, it has a more or less intensely green color. This color difference in the seed is clearly recognizable because the seed coats are translucent.
3. The *difference in color of the seed coat*: This is either white (combined with a white flower color), or, it is gray, grayish-brown or leather-brown, with or without violet spotting. In the latter case the color of

the standard is violet, that of the wings purple, and the stem at the leaf axils reddish. In boiling water, the green seed coats turn brownish black.

4. The *difference in the form of the ripe seed pod*: Either smoothly rounded without indentations, or deeply indented between the seeds and more or less wrinkly (*P. saccharatum*).
5. The *difference in color of the unripe pod*: Either light or dark green, or vividly yellow, which is also the color of the stem, ribs of the leaves, and the calyx.
6. The *difference in the position of the flower*: They are either axial, that is, distributed along the axis, or they are terminal, concentrated at the tip of the axis, and there occupy an almost pseudo-umbel-like distribution; in that case, a cross section of the upper portion of the stem shows it to be more or less widened (*P. umbellatum*).
7. The *difference in length of the axis*: The difference in length of the axis is quite different in individual forms; however, a constant trait for each insofar as it is subject to insignificant variations in healthy plants was raised in the same soil. To assure reliable differentiation in studies of this trait with a long axis of 180–210 cm was always crossed with a short one of 9–45 cm.

From a larger number of plants of the same type only the most vigorous were selected. Weak specimens always yield uncertain results because already in the first generation and even more so in the successive ones many offspring either do not flower at all or form few and defective seeds.

Furthermore, in all experiments reciprocal crosses were performed such that each of the two varieties that had served as seed plant in a number of prior crosses now served as the pollen plant.

The plants were raised in garden plots, a small part in pots keeping their naturally upright position with rods or tree branches connected with string.

For every experiment, a number of blooming pot plants were brought into the green house to serve as a control for the main experiment in the garden with reference to possible perturbations caused by insects. Of those which search out pea plants the beetle *Bruchus pisi* may impair the experiment, if in larger number. It is known that the female of this species opens the keel by depositing her eggs into the flower; under the magnifying glass, the tarsi of one specimen caught in a blossom clearly showed several pollen grains. It is necessary to mention here a circumstance, which may occasion the admixture of extraneous pollen. In a few rare cases, atrophy may occur in certain parts of the otherwise totally normally developed flower leading to a partial exposure of the generative organs. Thus, a defective development of the keel was observed whereby the stigma and anthers were partially uncovered. Also, occasionally the pollen does not attain its complete development. In such a case, during flowering, a gradual elongation of the pistil occurs until the stigma extends to the tip of the keel. This curious appearance was observed in hybrids of *Phaseolus* and of *Lathyrus*.

However, the risk of falsification by extraneous pollen is quite small in *Pisum* and not capable of distorting results at large. Among more than 10,000 plants examined in greater detail, it was the case only a few times that admixture could not be doubted. Such an interference was never observed in the greenhouse; this suggest that *Bruchus pisi* was responsible and perhaps also for the abovementioned abnormalities of the flower structure.

The form of the hybrids

The studies performed in former years on ornamental plants had shown that, as a rule, the hybrids did not represent an exact intermediate form between the parental types. Indeed with respect to a few more obvious characteristics such as shape and size of leaves, hairiness of individual parts, and so on, an intermediate form is almost always evident; however, in other cases one of the two major traits is so preponderant that it is difficult or completely impossible to detect the other in the hybrid.

This precisely is also the case in *Pisum*. Each of the seven hybrid forms is so completely like one of the parental types that the other disappears, or is so similar to it that a secure distinction is impossible. This circumstance is of great importance for the determination and the classification of the forms of the progeny of the hybrids. In the subsequent discussion, those traits which are transmitted into the hybrid union completely (or almost) unchanged and thus themselves represent the hybrid types, are designated as *dominant*, and those which become latent in the union as *recessive*. The term *recessive* was selected because the traits so designated withdraw or disappear completely in the hybrids but, as will be shown later, reappear unchanged in their progeny.

Furthermore, all experiments showed that it is totally irrelevant whether the dominant trait is characteristic of the seed or pollen plant; in both cases the hybrid form remains exactly the same. This interesting phenomenon is also emphasized by Gärtner with the comment that even the most experienced expert is incapable of deciding in a given hybrid, which of the two parental characteristics derived from the seed or the pollen plant.

Of the differential traits utilized in the experiments the following are dominant:

1. Round or rounded seed forms with or without shallow indentations.
2. The yellow pigmentation of the seed albumen.

3. The gray, grayish-brown, or leather-brown pigmentation of the seed cover in combination with reddish-violet blossoms and reddish spotting of the leaf joints.
4. The simple curved form of the pod.
5. The green pigmentation of the unripe pod in combination with the same color of the stem, the leaf ribs, and the calyx.
6. The distribution of the flowers along the stem.
7. The greater length of the stem.

With respect to the latter trait, it must be remarked that the longer of the two hybrid stems usually exceeds that of the tall parental stem, perhaps attributable to the great luxuriance of all parts of the plant when crossing plants with very different axes. So, for example, in repeated experiments involving axes of 30.5 and 183 cm the hybrid axes invariably ranged between 183 and 213 cm. The hybrids of the seed pods are frequently more spotted with some of the spots coalescing into smaller bluish-violet patches. The spotting appears frequently even when lacking in the parent.

The first generation of hybrids

In addition to the *dominant traits*, in this generation, the *recessive* ones appear in their full individuality and indeed in the definitely expressed average relationship of 3:1, so that among four plants in this generation three receive the dominant and one the recessive trait. Without exception this is true of all traits entered into the experiments. Without essential change there is reappearance in the numerical relationship of the angular wrinkled seed shape, the green pigmentation of the albumen, the white color of seed coat and of blossom, and the constrictions of the pods; the yellow color of the unripe pod, the stem, calyx, and ribs of leaves; the umbel-like arrangement of the flowers, and the dwarfed axis. *Transitional forms were not observed in any of the experiments.*

Since the hybrids resulting from reciprocal crosses had a complete form (i.e., were normal) and in their further development showed no remarkable aberration, it was possible to bring the results of each cross into a single calculation. The sums obtained for each of the two complementary traits are the following:

Experiment 1. Shape of seeds: In year two, 7,324 seeds were obtained from 253 hybrids. Of these, 5,474 were round or nearly round and 1,850 were angular-wrinkled. This yields the ratio 2.96:1.

Experiment 2. Pigmentation of the albumen: 258 plants yielded 8,023 seeds, 6,022 yellow and 2,001 green in a relationship of 3.01:1.

In both of these experiments, one usually obtains both kinds of seeds. In well-developed pods, which on the average contain 6–9 seeds, it was not uncommon that all seeds were round (experiment 1) or all yellow (experiment 2); in contrast, it was never observed that a pod contained more than five wrinkled or five green ones. It seems to make no difference whether the pod develops earlier or later on the hybrid, or originated from the main axis or a branch. In a few plants the earliest formed pods developed only single seeds and these then possessed exclusively one of the two characters; however, the relationship was normal in the later-developing pods. The distribution of characters in individual plans was as variable as in individual pods.

Both of these experiments are important for the determination of the averages because a small sample of experimental plants may yield very significant variations around averages. Thus, some care is required, especially in experiment 2, because in individual seeds of some plants the green color of the albumen is less developed and initially is easily overlooked. The cause of this partial disappearance of the green pigmentation bears no relationship to the hybrid nature of the plants because it also occurs in the parental plant; furthermore, this peculiarity is confined to the individual plant and is not hereditary in the progeny. This phenomenon is observed often in luxuriating plants. Seed damaged by insects during its development frequently vary in color and shape; however, with a little practice in sorting mistakes can be avoided. It is almost superfluous to mention that the pods have to remain on the plant until they have completely ripened and dried out, because only at that time shape and pigmentation are completely developed.

Experiment 3. Color of seed covers: Among 929 plants 705 had violet-red blossoms and gray-brown seed covers; 224 had white flowers and white seed covers. This gives a ratio of 3.15:1.

Experiment 4. Shape of pods: Of 1,181 plants, 882 [pods] were simply smooth, 299 had segmentally constricted pods. Hence, the ratio of 2.95:1.

Experiment 5. Pigmentation of the unripe pod: The total of experimental plants was 580, of those 428 possessed green and 152 yellow pods. Therefore, the former were in a ratio of 2.82:1 to the latter.

Experiment 6. Position of blossoms: Among 858 cases, the flowers were axial 651 times and terminal 207 times. Thus, the ratio of 3.14:1.

Experiment 7. Axial length: Of 1,064 plants, 787 had the long and 277 the short axis. Hence, the mutual relationship of 2.84:1. In this experiment, the dwarfed plants were carefully dug up and replanted into their own beds. This precaution was necessary because they would have perished through being overgrown by their tall siblings. Already in their earliest youth it was easy to recognize them because of diminished growth and the dark green thick leaves.

If now the results of all the experiments be brought together there is found, as between the number of forms with the dominant and recessive characters in an average ratio of 2.98:1 or 3:1.

[Mendel's paper goes on from this point for approximately 40 more pages developing his algebraic theory of inheritance.]

Epilogue four: What makes a good scientist?

To be still searching what we know not, by what we know,
Still closing up truth to truth as we find it, this is the
golden rule...

Areopagitica. John Milton

Scientists like to claim priority for any important new discovery. Newton and Leibniz for the elucidation of the calculus is a classic example. As a profession we are very good at awarding honors to the wrong person. Stigler has proposed a law claiming that no important scientific discovery is ever actually named after its original discoverer. Thus Newton's first law of motion was discovered by Galileo, Gay-Lussac's pressure–temperature law of gases was discovered by Boyle, the Gaussian distribution was used before him by de Moivre, and the famous Mobius strip was discovered shortly before Mobius by Johann Listing. To ignore the original discoverer is not necessarily due to plagiarism. If ideas are "in the air," the time may be ripe for several people to do the same decisive experiments. The person who makes the most noise about it often gets the credit. Darwin did not discover the theory of evolution; it was in the air at the time, thanks to Lamarck. Darwin, with the codiscoverer Alfred Wallace, proposed a plausible process of how evolution could be driven by natural selection. Mendel was entirely different. His 3:1 ratio was not "in the air" and he did not stumble on his discovery by accident. Many other scientists make their discoveries almost by chance. But Mendel worked his out laboriously over 8 years. Doing this is astonishingly difficult. It requires an enormous effort of will and concentration.

Chapter 6 describes how Francis Galton began a study on the qualities needed to become a successful scientist. He included Charles Darwin in the survey who when asked about special talents, replied that he had none! We expect a really good scientist to discover new facts about such important fields as heredity, cancer, and brain function that will open up new sets of ideas about how such systems actually work and eventually give us control over them. *In the field of observation chance favors only the prepared mind* wrote Louis Pasteur. Mendel needed to have more than intellect and concentration to achieve his results. His prepared mind was a love for his subject with a stunning inspiration that the language of heredity might be based on mathematics; but he had to wait for anything much to happen. He went on patiently with his life recording transient facts and the fugitive impressions of his plant hybrids, needing more than 8 years of investigation in his monastery, too impassioned to desist. And at the end of his studies there was no recognition or awards for his work from the outside world. This is an example of the purest science: to correctly predict the laws of nature, to confirm them by experiment and not care too much for the approbation of the rest of the world. Mendel's marvelous achievement is a particularly fine example that knowledge cannot spring from observation and experiment alone but by the comparison of ideas (or theories) with the observed facts. Seldom has anyone possessed such an intuitive grasp of hidden things combined with such a strong critical sense. With all his daily involvement in prayers, church services, Bible studies, and maintenance duties expected of a humble monk, he kept his eyes immovably fixed on the details of plant inheritance and the underlying principles that he intuitively perceived. Gregor Mendel is unquestionably one of the greatest discoverers of our age in the biological field. He defies almost all the canons by which we are accustomed to judge a scientist: a Catholic monk initially self taught in science, working without any professional colleagues in solitude in a monastery, and failing his major career examinations. The fact that he could express his discoveries in mathematics fills many of us with awe and wonder.

The activity of science is often portrayed as similar to Mendels' as a rather pure and abstract effort confined to the rarefied atmosphere of a laboratory (or monastery) to design experiments, to assemble the equipment, perform the experiments, and to analyze the results to see how they fit in with the existing body of knowledge.

The work of many scientists is not such a pure affair. At the beginning of a project, there can be confusion, frustration, rivalries, quarrels, and anger with no one knowing what things mean or in which direction to travel. Colleagues can become distrustful and do not stick to the original plan but go off at tangents; samples are labeled secretly; manuscripts are sent off for publication prematurely without discussion with the rest of the team. Galton at least had the decency to discuss his results with Darwin before publishing them.

The need to stick with a problem and work on it over many years against all the odds and many setbacks can lead to obsessional thinking—the picture of the mad scientist spending his life pursuing the crazy idea is a popular picture of our work. Part of the problem is to know when to give up the experiments testing the beloved hypothesis because too many facts have appeared to contradict it. This was Darwin's error with his gemmule hypothesis. Thomas Huxley drafted many aphorisms and one pointed out: *the great tragedy for a scientist is the slaying of his beautiful hypothesis by an ugly fact*. Huxley thought the pursuit of science as more like the royal game of chess; the chess board is the field of investigation we choose to work on; the movement of the chess pieces are the experiments we do. The player on the other side of the board is Mother Nature. We are obviously not trying to check mate our opponent, only trying to reveal more of the rules by which she works. We know some of the rules already but unlike the game of chess not all of them. Our opponent is external reality (or to keep sentimental, Mother Nature) and we know that her play is always fair, just, logical, and patient. However, we also know, to our cost, that she never overlooks a mistake, nor makes the smallest allowance for our ignorance. Our goal is to find out more of the essential rules of the game because we only have a partial knowledge of some. The outcome of a chess game is usually trivial with no major repercussions for the outside world; whereas there can be major implications resulting from a scientific enquiry, for example, the discovery of insulin by Banting and Best in 1921 that has saved the lives of millions of diabetic children.

There are other characteristics that good scientists need. Similar to a musician, poet, or painter a scientist is a discoverer of new patterns, made from the facts of nature, not of patterns made of words, sounds, or colors as for the artist (poet, musician, or painter). A great scientist is expected to discover new concepts about the world based on experimental evidence. The patterns established by the scientist are more permanent than those established by the painter or musician because the former always deals with the facts and ideas that correspond with the real world; whereas the artist seems to be equally concerned with our inner imaginative and subjective world. The patterns established by the mathematician are even more endurable, they hold for ever and are absolutely true; seven will always be a prime number whether the human or material world exists or not. Mathematicians would consider the scientists' results as more like approximations to reality.

Darwin once told Galton: *how odd it is that anyone should not see that all observation must be for or against some view if it is to be of any service.* And Thomas Huxley believed that every great step in science has been made by the anticipation of nature that is by the grasp of a hypothesis, which though verifiable, often has very little foundation in facts at the start. Without some guiding idea we do not know what facts to explore. The growth of our knowledge has to depend on new observations or new experiments. Darwin strongly maintained that the danger of false hypotheses is never as great as that of errors in observations. Hypotheses are similar to fishing nets; you have to cast out before you can catch anything. A bad hypothesis may still enable us to advance the organization of our material even if catching nothing with this particular net. However, everything that is based on false observation has to be undone before we can build anew on more reliable facts.

A good original idea is only half the battle. The other half is to demonstrate that it is true and important. Defining truth is a complex issue for the philosopher, but for the scientist it can mean that the observations can be replicated by anyone under the appropriate conditions. Defining importance likewise can mean many things to different people, but for the scientist it can mean that the observations are unequivocal, unexpected, have generality, have depth, and have a practical impact. Thus Mendel's results were completely unexpected, they were easy to reproduce after choosing the appropriate conditions, they apply across a whole range of sexual reproduction in plants, insects, and animals, they led to the discovery of the chemical nature of the gene and DNA, as well as being of practical value in so many different ways (e.g., synthesizing biological molecules that act as pharmacological agents, predicting the occurrence of inherited disease in humans).

The interpretation of scientific experiments demands a total respect for the factual evidence obtained; however, much of the facts go against one's deepest wishes to establish the validity of the original idea. The temptations to modify the results; to discard results that do not fit the idea; to simply ignore some of the adverse data; to be unwilling to correct inaccurate information if the latter supports the hypothesis; or to modify the interpretation of the results to fit one's pet theory can lead to an immense waste of time and resources.

Personal vanity is assuredly a most common and disadvantageous quality for a scientist, because it leads to self-satisfaction, loss of adaptability, and a decline in imagination. A rigidity of spirit can deprive the ageing scientist of his adaptability and humor. He fails to respond with his former elasticity to ideas and theories of which he disapproves or to new developments with which he is not familiar. The senior scientist, as he loses the gift of imagination, becomes all keel and ballast without sail. He fails to react adequately to new winds that blow from different quarters; or to the squalls that arise within his own laboratory. Some scientists have a tendency to get into ruts, their thinking becomes ossified, and they apply their ideas or techniques regardless of the available evidence or to varying circumstances.

From the start, Darwin and Galton always professed to be scientific men, even in those early days of South American and African explorations. However, the wives of Darwin and Galton took different views upsetting their husbands in the process. They had both been brought up in the Christian faith and Emma Darwin confessed that her husband's evolutionary views made her feel quite desolate, particularly, if they included spirituality and the subject of morality. She admonished her husband that he might not have taken all the pains he should have done to judge truly on such important matters as spirituality and religion, which are clearly not science, nor amenable to scientific method. Louisa Galton took similar views. They both seemed to feel strongly that there were serious limitations to the scientific method and the knowledge so gained. Can one maintain that scientific research is superior to all other pathways to knowledge? Some branches of knowledge, such as religion or the ancient myths, do not conform to even the most straightforward rules of logic yet still have important messages to impart. How secure is a philosophy of life that is based solely on the results of laboratory measurements? A priest, like Mendel, is a man using a divining rod; a scientist, like Mendel, is a man using a microscope and test tube. Happy is the man who can integrate the two. The Psalms may seem to be illogical but are still an inspiration to many people. There may be a different sort of process working here that is more inspirational than intellectual; religion can energize poets, painters, musicians—and scientists like Mendel.

But, the reader may object, "to do science successfully you have forgotten to mention patience, discernment, courage, and the tact and charm that make a good team worker." No—I have just taken the presence of these for granted.

References, notes, and a brief gene timeline

This book is a popular account of the subject and some readers believe that footnotes and references disrupt the pleasant progress of a text and are a sign of pedantry or an insufficiently realized task. Others of a more scholarly bent want chapter and verse quoted for every sentence that contains a fact. I have tried to steer a middle course and give the sources of information for each chapter at the end. Those who are annoyed by endnotes, postscripts, and provisos can perhaps take my probity on faith and disregard the notes that follow.

Bibliography

Darwin

Appleman P. *Darwin: A Norton Critical Edition*. W.W. Norton & Co., New York, 1970.

Browne J. *The Power of Place*. Vol. 11 of Darwin's biography. Jonathan Cape, London, 2000.

Burkhardt F et al. (Eds). *The Darwin Correspondence Project*. Cambridge University Press, Cambridge, UK, 1985 - ongoing. This ambitious project is to publish all Darwin's letters in about 30 volumes. Volumes 1–21 are already published and the library gave me access to unpublished letters from 1869 onwards.

Burkhardt F. *Charles Darwin's Letters: A Selection 1825–1859*. Cambridge University Press, Cambridge, UK, 1996.

Darwin C. *Autobiography (with essays on everyday life and religion by F Darwin)*. Watts & Co., London, 1937.

Darwin C. *Journal of Researches*. Nelson & Sons, London, 1903.

Darwin C. *The Descent of Man and Selection in Relation to Sex*. Princeton University Press, Princeton, NJ, 1981.

Darwin C. *The Origin of Species by means of Natural Selection, or the Preservation of Favoured Races in the Struggle for Life*. John Murray, London, UK, 1901.

Darwin C. *The Variation of Animals and Plants under Domestication*. Volumes 1–2. John Hopkins University Press, Baltimore, MD, 1998.

Darwin F. *The Autobiography of Charles Darwin and Selected Letters*. Dover Publications, New York, 1958.

Darwin C, Wallace AR. On the tendency of species to form varieties; and on the perpetuation of varieties and species by natural means of selection. *J. Proc. Linn. Soc. Lond.* 1858, 3:45–62.

Desmond A, Moore J. *Darwin.* Penguin Books, London, 1992.

Healey E. *Emma Darwin: The Inspirational Wife of a Genius.* Headline, London, UK, 2001.

Miller J, van Loon B. *Darwin for Beginners.* Writers & Readers Publishing Society, London, UK, 1982.

Platt R. Darwin, Mendel and Galton. *Med. Hist.* 1959, 3(2):87–99.

Raverat G. *Period Piece.* Norton & Co., New York, 1953.

Ritchie D. *Darwinism and Politics.* Swan Sonnenschein & Co., London, UK, 1889.

Galton

Allen D. *The Forgotten Brummie. The Life and Legacies of Sir Francis Galton.* Charleston Press, Charleston, SC, 2014.

Forrest DW. *Francis Galton: The Life and Work of a Victorian Genius.* Paul Elek, London, UK, 1974.

Galton F. *Essays in Eugenics.* Scott-Townsend, Washington, DC, 1996.

Galton F. *Hereditary Genius.* Macmillan & Co., London, UK, 2nd ed., 1892.

Galton F. *Inquiries into Human Faculty & Its Development.* J.M. Dent, London, UK, 1883.

Galton F. *Memories of My Life.* Methuen & Co., London, UK, 1908.

Galton F. *Natural Inheritance.* Macmillan & Co., London, UK, 1889.

Galton F. *The Art of Travel, or Shifts and Contrivances Available in Wild Countries.* Phoenix Press, London, UK, 1971.

Galton F. *Travels of an Explorer in Tropical South Africa.* Ward, Lock & Co., London, 1890.

Galton L. Unpublished diaries. University College Library, London, UK.

Gillham NW. *A Life of Sir Francis Galton.* Oxford University Press, New York, 2001.

Keynes M (Ed). *Sir Francis Galton FRS: The Legacy of His Ideas.* Macmillan Press, London, UK, 1993.

Pearson K. *Life, Letters and Labours of Francis Galton.* Volumes 1–111b, Cambridge University Press, Cambridge, UK, 1914.

Galton and Eugenics

Armstrong C. Thousands of women sterilized in Sweden without consent. *Brit. Med. J.* 1997, 315:563.

Coghlan A. Saviour sibling babies get green light. *New Scientist* 2004, 13:28.

Convention on Human Rights and Biomedicine. European Treaty Series 164, Oviedo 1997. Publ. Edition du Conseil de l'Europe.

Convention on the Rights of the Child. Publ. Editions du Conseil de l'Europe, 1989.

Editorial: A role model of rigidity. *Lancet* 1996, 348:1253–1254.

Galton DJ. *Eugenics, the Future of Human Life in the 21st Century.* Abacus, London, UK, 2001.

Galton DJ, O'Donovan K. Legislating for the new predictive genetics. *Hum. Reprod. Genet. Ethics* 2000, 6:39–48.

Kevles DJ. *In the Name of Eugenics.* Harvard University Press, Cambridge, MA, 1995.

The Ethical Aspects of Prenatal Diagnosis. European Commission of Advisors on the Ethical Implications of Biotechnology, Brussels, Belgium, 1996.

Mendel

Bateson W. *Mendel's Principles of Heredity.* Cambridge University Press, Cambridge, UK, 1902.

Centenary of Gregor Mendel and of Francis Galton. The Scientific Monthly, March 1923. Collection of papers presented at the American Society of Naturalists, Boston, MA, December 29, 1922.

Curt S, Sherwood ER. *The Origin of Genetics: A Mendel Source Book.* W. H. Freeman, San Francisco, CA, 1966.

de Vries H. Sur la loi de disjonction des hybrides. *Compte Rendus* 1900, 130:845–847.

Fisher RA. Has Mendel's work been rediscovered. *Ann. Sci.* 1936, 1:115–137.

Franklin A, Edwards AWF, Fairbanks DJ, Hartl DL, Seidenfeld T. *Ending the Mendel-Fisher Controversy.* University of Pittsburgh Press, Pittsburgh, PA, 2008.

Galton DJ. Did Mendel falsify his data? *Q. J. Med.* 2013, 105:215–216.

Henig RM. *A Monk and Two Peas.* Phoenix, London, 2001.

Iltis H. *Life of Mendel.* Norton, New York, 1932.

Keynes M, Edwards AWF, Peel R (Eds). *A Century of Mendelism in Human Genetics.* CRC Press, Boca Raton, FL, 2004.

Mendel G. Versuche uber Pflanzen-Hybriden. Verh naturf. Ver in Brünn; band iv 1865. English translation in Journal. Roy. Hort. Soc. 1901 xxvi.

Mendel GJ. Ueber einige aus kunstlicher Befruchtung gewonnen Hieracium-Bastarde ibid viii, 1869.

Moore R. The re-discovery of Mendel's work. *Bioscene* 2001, 27:14–22.

Olby RC. *Origins of Mendelism.* University of Chicago Press, Chicago, IL, 1985.

Opitz JM, Pavone J, Corsello G. The power of stories in pediatrics and genetics. *Ital. J. Pediatr.* 2016, 42:35–46.

Orel V. *Gregor Mendel: The First Geneticist.* Oxford University Press, Oxford, UK, 1996.

Other characters

Bearn AG. *Archibald Garrod and the Individuality of Man.* Clarendon Press, Oxford, UK, 1993.

Cadbury D. *The Dinosaur Hunters.* Fourth Estate, London, 2001.

Desmond A. *Huxley: From Devil's Disciple to Evolution's High Priest.* Addison-Wesley, Reading, MA, 1997.

Fisher RA. *The Genetical Theory of Natural Selection.* Dover Publications, New York, 1958.

Harris H. *Garrod's Inborn Errors of Metabolism.* Oxford University Press, Oxford, UK, 1993.

Huxley TH. *Man's Place in Nature and Other Essays.* J. M. Dent, London, 1906.

Opitz JM, Pavone L, Corsello G. The power of stories in paediatrics and genetics. *Ital. J. Pediatr.* 2016, 42:35.

Owen R. *Life of Professor Owen*. Volumes 1–2. John Murray, London, UK, 1895.
Romanes M. *Life and Letters of George John Romanes*. Longmans, Green & Co., London, UK, 1902. (written by his wife after his death).
Strachey L. *Eminent Victorians*. Chatto & Windus, London, 1929.
Sturtevant AH. *A History of Genetics*. Harper & Row, New York, 1965.
Turrill WB. *Joseph Dalton Hooker*. Scientific Book Club, London, UK, 1963.
Wallace AR. *My Life: A Record of Events and Opinions*. Volumes 1–2. Chapman & Hall, London, UK, 1905.

Scientific method: there are hundreds of books on this topic. I have found the following useful

Ayer AJ. *The Foundation of Empirical Knowledge*. Macmillan & Co., London, UK, 1964.
Ayer AJ. *The Problem of Knowledge*. Penguin Books, London, UK, 1956.
Chalmers AF. *What is This Thing called Science?* Open University Press, Milton Keynes, UK, 1985.
Cohen MR, Nagel E. *An Introduction to Logic and Scientific Method*. Routledge & Kegan Paul, London, UK, 1961.
Polya G. *How to Solve It*. Princeton University Press, Princeton, NJ, 1973.
Popper KR. *The Logic of Scientific Discovery*. Harper & Row, New York, 1965.

Topic references: Darwinians v Mendelians

Bateson B. *William Bateson FRS, Naturalist*. Cambridge University Press, Cambridge, UK, 1928.
Bateson W. *Mendel's Principles of Heredity*. Cambridge University Press, Cambridge, UK, 1902.
Bulmer M. Galton's law of ancestral heredity. *Heredity* 1998, 81:579–585.
Garrod AE. The incidence of alkaptonuria: A study in chemical individuality. *Lancet* 1902, 2:1616–1620.
Pearson K. Mathematical contributions to the theory of evolution: On the law of ancestral heredity. *Proc. R. Soc.* 1898, 62:386–412.
Zoology at the British Association. *Nature* 1904, 70:538–541; *Times Saturday* August 20, 1904.

Garrod's Inborn Errors of Metabolism

Garrod AE. A contribution to the study of alkaptonuria. *Med-chir Trans*. 1899, 82:369–394.
Garrod AE. The incidence of alkaptonuria: A study in chemical individuality. *Lancet* 1902, 2:1616–1620.
Garrod AE. The Croonian lectures on inborn errors of metabolism. *Lancet* 1908, ii:1–7; 73–79; 142–148; 214–220.
Galton DJ. Archibald Garrod (1857–1936). *J. Inherit. Metab. Dis.* 31:561–566.
Galton DJ. Archibald Garrod: The founding father of biochemical genetics. In: *Pioneers of Medicine without a Nobel Prize*, G. Thompson (Ed.). Imperial College Press, London, UK, 2014.

Scriver CR. Garrod's Croonian lectures and the charter 'inborn errors of metabolism'. *J. Inherit. Metab. Dis.* 2008, 31:580–598.

Weatherall DJ. The centenary of Garrod's Croonian lectures. *Clin. Med.* 2008, 8:309–311.

Notes

[1] seq. Boyhood: The details of Galton's childhood and boyhood are taken from his autobiography (Galton 1908) and his biography ref. Darwin's autobiography (Darwin 1937) does not tell about his childhood but by 1827 he was writing letters that reveal his opinions. The episode of the assassin bug taken from (Darwin 1903).

[2] Galton's will made at the tender age of 8 years is published in Pearson (1914, volume 1).

[3] Galton's beauty map was made with the aid of small pocket registrator. This was a metal instrument specially made for him by Hawksley. It consisted of a base on which five dials were attached. The dials record the number of separate pressures exerted on five stops that communicate by a ratchet with a separate index arm that moves round its own dial. The instrument is ¼ inches thick, 4 inches long, and 3/4 inches wide. Guides are placed to keep the fingers in their proper places on the stops. With this instrument in his pocket he recorded the numbers of very attractive, attractive, indifferent, and repellent looking women that he met on his walks through the streets of various towns to construct the map. He observed London to have the most and Aberdeen the fewest numbers of attractive women.

Another scientific instrument he invented was the Galton whistle. This was an instrument to test the limits of audibility of a person. Galton observed that the pitch of a note produced by a cylindrical whistle depend on its length. Therefore, he altered the length of a tube whistle by a screw-in plug at the closed end. The numbers of turns are registered on a scale fixed to the barrel of the whistle. The pitch of the screw was 25 to the inch. One turn of the screw shortens the tube therefore by 1/25th of an inch. With this instrument, he found that older people have a high-pitch hearing loss compared to younger subjects.

He also devised a machine in 1889 to simulate the binomial theorem as used by Mendel. He used a jar into which he poured lead shot through a funnel. A wedge (A) placed symmetrically below the funnel split the stream of lead shot into two; below this were two more wedges (B) and (C) that split the stream of shot symmetrically again and he collected the final stream of lead shot into three compartments at the bottom of the jar. Counting the lead shot in each compartment he

found it approximated to the binomial probability of: $P(X = r) = nCr.$
$p^r. (1 - p)^{n-r}$ for $0 < r < n$; in contrast the binomial theorem reads:
$(a + b)^n = nCr.a^{n-r}.b^r$.

[4] seq. Medicine. Darwin gave his own account of his medical studies
in his letters (Burkhardt et al. 1985; Burkhardt 1996) and in his autobiography (Darwin 1937). Galton's account of his medical studies is
in Forrest (1974) and Galton (1908).

[5] seq. Explorations. Darwin (1903) is the famous account of *Darwin's
travels to South America*; later called *the Voyage of the Beagle*. Galton's
travels are written up in Galton (1890, 1971) that is surprisingly still
in print.

[6] seq. Regarding all the letters in the text, the ones relating to Darwin
come from Burkhardt et al. (1985), relating to Galton from Pearson
(1914), to Huxley (Desmond 1997) and to Romanes (1902).

[7] seq. To be a husband. I have taken most of the account of Darwin's
courtship and married life with Emma Darwin from Healey (2001).
There are also many letters between the two in Burkhardt et al. (1985).
Galton's courtship and married life are taken from his autobiography
(Galton 1908) and Louisa's diary (Galton, Unpublished diaries) with
Pearson's biography (Pearson 1914). However one has the impression that Pearson disapproved of Louisa Galton since there are many
details one would expect a dispassionate biographer to report that
are not included.

[8] Galton's probable venereal infection is documented in Pearson (1914)
and Gillham (2001).

[9] Two books. So much has been written about The *Origin of Species* that
an author is spoilt for choice of sources. Appleman (1970) is broad
ranging and Darwin (1958) gives the Darwin family view of the book.
Wallace's account of his own independent discovery of natural selection is taken from Wallace (1905).

[10] Church Militant against Darwin: science became unafraid to contradict theological dogma from the start of the nineteenth century even
at the risk of offending many powerful people, especially those who
tried to impose their ideas that were clearly shown to be incorrect
by the facts of natural history, such as the age of the Earth. Scientists
demanded that knowledge be gained to the greatest extent by scientific methods using the best available equipment of the time. See
Epilogue 1 for the Churchs' views against science.

[11] seq. The details of Huxley's life come from the extensive biography
by Desmond (1997) and the interaction between Darwin and Huxley
from the letters (Burkhardt et al. 1985; Appleman 1970). The major
source for Hooker was the biography by Turrill (1963) and other
details have come from the recent Darwin biographies (Browne 2000;
Desmond and Moore 1992).

[12] Galton's Hereditary Genius is his most famous book (Galton 1892); unlike Origin of Species is not in print now. The chapter on General Considerations gives a complete account of Darwin's gemmule hypothesis that remains in the second edition of 1892. That is 21 years after Galton had published experiments disproving (to his mind) the gemmule hypothesis.

[13] seq. Grand theory. The letters related to the gemmule hypothesis come from Burkhardt et al. (1985) and the hypothesis is formally presented in Darwin (1998). The details of Hooker's and Huxley's opposition to the hypothesis are taken from letters in Burkhardt et al. (1985) and the biographies of Darwin (Browne 2000; Desmond and Moore 1992). Browne (2000) is a professional historian's account of the topic for more detail.

[14] An entertaining account of the Darwin household is given by his granddaughter in Raverat (1953).

[15] seq. Trial by Experiment. The whole of the Galton–Darwin episode relating to the rabbit experiments is taken from the Pearson biography (Pearson 1914). Many of the extant letters are published here. This biography is a veritable mine of information about Galton's relationship with Darwin including a wide selection of letters between the two and accounts of their meetings. Despite the masses of material, the volumes are extremely well indexed thereby facilitating the search for particular facts.

They are also described in the Darwin biographies (Browne 2000; Desmond and Moore 1992) but given much less space. Galton also gives an account of the episode in his autobiography (Galton 1908). The account is simplified for the benefit of the general reader who may not be interested in the sexual habits of rabbits.

[16] Brünn in Moravia is now called Brno and is part of the Czech Republic.

[17] The Fibonacci number sequence found in biology (number of leaves on a stem or number of petals on a flower) relate to some aspects of genetics. The sequence reads: 1,1,2,3,5,8,13 where each number after the second is the sum of the two previous numbers. The sequence of the adjacent number ratios at a limit is the Golden Ratio.

[18] The Chi^2 probability test shows that Darwin's results fitted the Mendelian ratio of 1:3 quite well (giving a Chi^2 value of 0.25 for rejecting a real difference between the two ratios—it should be less than 0.05).

[19] Mendel's reprint. The most contentious issue in this chapter deals with the question whether Darwin received a copy of Mendel's article? Reams have been written on the subject by historians and Darwin scholars alike. To summarize the arguments:

Against Receipt. A catalogue of Darwin's library from Down House published in 1908 (26 years after Darwin's death) did not record any of Mendel's papers. After Darwin's death in 1882 his scientific library passed to his son Francis. Down House was cleared of its contents in 1896 following the death of Emma Darwin and the house then leased to a school. Francis Darwin later bequeathed the book collection to the Professor of Botany at Cambridge University and a catalogue of the library was prepared by H.W. Rutherford and published in 1908. There was thus ample time for items to go astray.

The catalog did record the presence of Focke's and Hoffmann's books; and the former mentions Mendel's claim to have found "constant numerical relationships" among the different phenotypes of the F2 generations after plant hybridizations.

For receipt

1. Mendel ordered forty reprints of his paper of 1866 to send to famous European scientists; and Darwin by then was certainly one of the most famous. Mendel also sent copies to Learned Societies such as the Linnaean and Royal Societies. Darwin's book on *Origin of Species* had been out for 6 years and was already in its third edition. It had been translated into German, French, Dutch, Spanish, Polish, and Russian. By 1876, it had sold 16,000 copies, which is a large sale for such a "stiff" book.

2. Mendel had studied Darwin's book (the 1863 German edition), annotated it in handwriting and mentioned in his own paper the implications of his work for the evolution of species. Darwin was therefore likely to be one of the recipients.

3. Darwin was not sympathetic to a mathematical presentation of data and Mendel's paper was full of algebraic expressions. If one cannot understand a paper one tends to ignore it.

4. Several biographers and scholars have written that Darwin was sent a copy of Mendel's paper. For example R. Henig mentions it in Henig (2001) and says that she obtained her information about this from Anna Matlova, who was director of the previously named Mendel Museum in Brünn. Of course sending a reprint is not the same as receiving one. Other writers who maintain that a paper was sent (and received) are P. Kitcher in *Abusing Science*; M. Rose in *Darwin's Spectre* and G. Dover in *Dear Mr. Darwin*. For a scholars account of the uptake of Mendel's work before 1900. see R. Olby and P. Gantrey 1968 *Eleven references to Mendel before 1900* in Annals of Science 24:7–20. On balance my own view would be that it was 99% likely that Darwin was sent the crucial paper by Mendel, but even if read would probably not change his opinions.

[20] seq. The quarrel. The two men were corresponding actively during the time of the rabbit experiments considering the results and exchanging experimental material. So it is unlikely that they did not meet in person for a discussion since Galton took Darwin to be a co-investigator. One would assume the drift of the discussion to be on the topics as described.

[21] The idea of genetic factors in the blood determining inheritance persisted well after Darwin's death. The first Nuremberg Race Law of 1935 was for *Protection of German Blood* by prohibiting marriage between Jews and Germans. German blood might be contaminated by such a union. *The Reich Citizenship Law* declared those not of German blood to be state subjects, whereas those classified as Aryans were citizens of the Reich. The idea of genetic factors in blood has been reborn with the recent observations that DNA fragments can be found in the blood stream of patients with cancer. The DNA may come from damaged cancer cells or perhaps be involved with their spread.

[22] seq. Galton–Darwin Publications: these excerpts are taken directly from published articles by Galton and Darwin. The first draft of Galton's reply where he compares Darwin to an ape-like animal appears to be lost but the letter in *Nature* April volume of 1871 is as published.

[23] The referee. Romanes's wife wrote a very detailed biography (Romanes 1902), no doubt as a memorial for her husband's premature death in 1894.

His prosperous Victorian life-style is well described and I have used no further sources.

[24] As Max Planck (the Nobel prize winner for physics in 1918) was to write many years later: *A new scientific truth does not triumph by convincing its opponents and making them see the light; but rather because its opponents eventually die and a new generation grows up that is familiar with it.*

[25] Galton's modifications of his views on heredity from gemmules to germs and stirps are well described in Forrest (1974) and Gillham (2001).

[26] seq Greenhouse at Down. The potato experiments are to be found in Romanes' biography (1902); as is the description of his lectures in Glasgow, Leeds, and Birmingham. His poetic activities are to be found in Romanes' biography (1902).

[27] The Linnaean Society in Victorian England: membership of such societies (the Linnaean, British Association for Advancement of Science, Meteorological, Royal Horticultural, and the Royal Society) carried as much prestige as holding a university post. Universities

often evolved from teachers instructing private pupils in their own homes banding together to form educational societies. The oldest university in England may have been Cambridge that began in 1209 when a number of disaffected students from Oxford formed a "society" for higher education. The oldest Royal Society in England is the College of Physicians (1518), the Royal Society (1660), Botanic (Kew 1759), and Geographical (1830). The importance of these societies diminished in the twentieth century due to the development of the university system.

[28] Wallace's change of views on heredity is to be found in His Life (Wallace 1905).

[29] seq. Darwin's Death: This account is an amalgamation of details from Romanes (1902), Desmond and Moore (1992), and Appleman (1970). The ode to Darwin is published in Romanes (1902).

[30] Romanes death: This is described movingly by his wife in Romanes (1902). One cannot help admiring his evident courage in the face of a premature death. His book *Darwin, and after Darwin* was published posthumously by Longmans, Green & co. 1900.

[31] ... work on Darwin's gemmule hypothesis. It seems that many men take the limits of their own field of vision for the limits of the whole problem.

[32] Galton's travels to Egypt; to travel 14 miles on a donkey in Egypt at the age of 78 years to visit the ruins testifies to Galton's vigor and lively interest in things. The details of these travels mainly come from letters published in Pearson (1914).

[33] *Mendel...solitary in his monastery*: In Galton's lecture of 1905 he was still designing experiments to test the applicability of the Mendelian hypothesis in man (communicated at a meeting of the School of Economics and Political Science, Clare Market, on Tuesday February 14, 1905 at 4.00 p.m.).

[34] Bateson: I am indebted primarily to three accounts of Bateson in Gillham (2001) and Bateson (1902, 1928). There are more accounts of his life in Bateson (1928) written by his wife.

[35] "Truth" for Galton or Mendel is not easy to define. It can be taken to mean an accurate correspondence of our ideas with an external reality combined with a logical coherence within our currently held set of ideas and beliefs. On this basis, it is always approximate, because scientific studies are always improving the degree of correspondence and degree of coherence with reality. It is quite different from the truth of Mendel's algebraic propositions that usually come down to a question of logic. For a Man of God like Mendel religious truth is mainly based on revelation and conveys nothing clear to many

scientists. It starts from some irrational source, such as revelation or arbitrary authority (e.g., *God created Man*) and then elaborates on a number of possible consequences. However, it can never change the initial starting point and all subsequent facts must be adapted to it. Scientists on the other hand never accept an immutable starting point; they test their axioms continually by experiment, and when found wanting the axioms are discarded without hesitation. Mendel seems mentally to be able to hold these three separate definitions of truth concurrently.

[36] Garrod: I could only find one biography of Garrod (Bearn 1993) from where all the family details are taken. His scientific achievements are given full credit in Harris (1993) and in a chapter by me in *Pioneers of Medicine* without a Nobel Prize 2014, p. 1–21 published by Imperial College Press.

[37] Botanists rediscovering Mendel. Much has been written on the topic of why Mendel was neglected for so long (Centenary of Gregor Mendel and of Francis Galton [1923] gives a detailed assessment). The papers by de Vries and Correns rediscovering Mendel works in 1900 are available on the web.

[38] Mendelians v Darwinians. This acrimonious dispute of 1904 between the Mendelians represented by Bateson, and the Darwinians represented by Pearson and Weldon is well described in Gillham (2001).

[39] Pearson refined Galton's Law of Ancestral Inheritance to read:

$$\text{Inheritance} \sim (0.5)\, b_1 k_1 + (0.5)^2 b_2 k_2 \ldots + (0.5)^n b_n k_n$$

where $b_1, b_2 \ldots b_n$ are the ratios of the standard deviations of traits of offspring to the standard deviations of the mid-parental generation (σ_0/σ_n in Pearson's notation); and $k_1, k_2 \ldots k_n$ are the deviations of the mean mid-parents from the mean of the offspring.

[40] A Taylor convergent series is of the general form:

$$F(x) = A + Bx + Cx^2 + Dx^3 + \ldots$$

where A, B, C, D… are constants independent of x (=0.5 in Pearson's expression). The expression can be used to represent a great many biological phenomena. The formulation has no theoretical significance; all it postulates is that the phenomenon in question varies continuously. Then Maclaurin's or Taylor's theorems (depending on the number of variables involved) can be used to determine the values of the coefficients that will make the series useful to any desired degree of approximation.

[41] Weldon rolled a set of 12 dice 26,306 times to judge whether the differences between a series of group frequencies and a theoretical law were or were not more than might be attributed to the chance fluctuations of random sampling. Weldon's dice data were used by Karl Pearson in his pioneering paper on the Chi-squared statistic.

[42] "Well and truly on the fence": Galton published many articles on inheritance starting in 1865 with one entitled Hereditary Talent—the same year that Mendel communicated his studies on plant hybrids to the Scientific Society at Brünn. Then followed other publications: A Theory of Heredity (1875); Heredity in Twins (1876); Typical Laws of Heredity (1877); Chance and Its Bearing on Heredity (1886); Family Likeness in Eye Color (1886); and the last paper on heredity in 1889 on the Basset hounds. He also published two books on the subject, Heredity Genius (1869) and Natural Inheritance (1889). His study on eye color again supported Mendel's views. His observations were made on 948 children, 336 parents, and 449 grandparents. He found that eye color generally bred true. Moreover, if one parent had a light eye color and the other a dark eye color, a few of the children were light and the rest dark. There were seldom children with intermediate eye colors, which would have been expected if blending inheritance had occurred.

When Mendel's views started to gain ground after 1900 Galton's own work on heredity seemed to lose momentum and he became active in a different field from 1901 onward with eugenics, publishing such articles as: "Improvement of the Human Breed" (1901); "Eugenics, its Definition, Scope and Aims" (1905); "Eugenics as a Factor in Religion" (1905); "Probability—the Foundation of Eugenics" (1907); "An Address on Eugenics" (1908); "Eugenic Qualities of Primary Importance" (1910). A change of field is not uncommon among scientists if they feel they have made a mess of their previous investigations.

[43] See Galton (2012), attempts to rectify this undeserved attack by Fisher and the book Franklin et al. (2008). They concluded that Mendel was not guilty of fraud.

[44] Parent plants produce the first generation of offspring plants called the F1 generation; they in turn produce after self- or cross-fertilization, the second generation of offspring called the F2 generation and so on. F stands for filial.

[45] G. H. Hardy (1877–1947) the Cambridge mathematician who was not interested in applying mathematics to genetics (or to anything else) with Weinberg formulated the expression where (p) and (q) now represent the allele frequencies in a large, randomly mating population, not undergoing selection and allows calculation of the frequencies of expected genotypes. It represents the probability of combined events and so is multiplicative, for example,

$$p^2 + 2pq + q^2 = 1$$

[46] Kepler's two planetary laws were invaluable to Newton:
(1) Planets move in ellipses round the Sun; (2) a line drawn from the center of the Sun to the center of the planet will sweep out equal areas in equal time intervals. Kepler's planetary measurements were of equal assistance for Newton.

A brief gene timeline*

*1859: Darwin publishes *On the Origin of Species by Means of Natural Selection*. He needed an explanation for the inheritance of beneficial variants for natural selection to operate.

*1866: Mendel founded a theory of heredity as the discrete transmission of *unit factors* (later called genes) in the edible pea determining seven different characteristics such as seed color, plant height, flower color, and so on. It was published in the proceedings of the Society for the Study of Natural Science in Brünn.

*1868: Darwin publishes *Variation of Animals and Plants under Domestication* that expounds his ideas on blending inheritance by way of pangenesis and gemmules.

*1869: Galton publishes his book *Hereditary Genius.* Its introduction supports Darwin's ideas on blending inheritance, although his own work tends to disprove it. Further editions still credit Darwin's theory of inheritance.

J. F. Miescher discovers nucleic acids (DNA) in cells and calls it *nuclein*.

1878: Flemming describes chromosome, later shown to carry the genes. He stains chromosomes with dyes to observe them clearly and describes them during the whole process of mitosis in 1882.

*1900–1902: Three botanists de Vries, Correns, and Tschermak independently rediscover Mendel's ratios while doing their own studies on inheritance in various plants. Garrod observes Mendelian ratios for the inheritance of some rare human diseases.

*1908: Garrod shows that the inherited disease alkaptonuria (*black urine* disease) is caused by a defective enzyme on a metabolic pathway; thus, linking Mendelian unit factors to enzymes and biochemistry.

1909: Wilhelm Johannsen coins the word *gene* in his book *Elemente der exakten Erblichkeitslehre.* (publ. Gustav Fischer, Jena) to describe the Mendelian unit of heredity. He was the professor of plant physiology and genetics in Copenhagen, Denmark.

* Topics covered in the book.

1909 onward: Thomas Hunt Morgan and his students examined mutations of fruit flies, and related them to positions on chromosomes by genetic linkage analysis.

1941: Tatum and Beadle replicate Garrod's work of 1908 showing that one gene can go to form one enzyme.

1944: Avery, Macleod, and McCarty identify the transforming substance of Fred Griffith's work to be DNA that can alter the inherited structure of bacterial cell walls.

1952: Chargaff discovers that DNA from many cells has a 1:1 ratio of pyrimidine to purine nucleotides; or more specifically the amount of guanine is equal to the amount of cytosine, and the amount adenine is equal to thymidine (the base pair rule).

1953: Crick, Watson, Wilkins, and Franklin work out the double helical structure of DNA using X-ray crystallography.

1961–1965: The genetic code is cracked by Nirenberg and others as a four-letter alphabet in which three letters (triplets) determine the order of 20 kinds of amino acids in proteins.

2003: The Human Genome Project reports the sequence of more than 20,500 human genes (at least 99.3% accurate).

2017 onward: It will take decades of more research for scientists to understand all of the information that is contained within the human genome. In time, more human diseases will be understood at the level of the molecules that are involved, which could fundamentally change the practice of medicine by leading to the development of new drugs, as well as to genetic testing to improve and personalize medical treatments.

Credits

Epigraphs

Permission to use epigraphs from the following poems and prose have been granted by: Faber & Faber Ltd. *"Blessed is the Man"* from selected poems by Marianne Moore; J. M. Dent Everymans, Paradise Lost, and Areopagitica, John Milton; Harvard University Press extracts of poems on pages 55 and 176 reprinted by permission of the publishers and the Trustees of Amherst College from THE POEMS OF EMILY DICKINSON, edited by Thomas H. Johnson, Cambridge, Mass. The Belknap Press of Harvard University Press, Copyright © 1951, 1955 by the President and Fellows of Harvard College. Copyright © renewed 1979, 1983 by the President and Fellows of Harvard College. Penguin Poets. Satire Three on Religion. John Donne; and Quarterly Journal of Medicine (OUP) CODA articles in volumes 102, 104, and 105.

Index

Note: Page numbers followed by f refer to figures in the text.